NASA STI Program ... in Profile

Since its founding, NASA has been dedicated to the advancement of aeronautics and space science. The NASA scientific and technical information (STI) program plays a key part in helping NASA maintain this important role.

The NASA STI program operates under the auspices of the Agency Chief Information Officer. It collects, organizes, provides for archiving, and disseminates NASA's STI. The NASA STI program provides access to the NASA Aeronautics and Space Database and its public interface, the NASA Technical Report Server, thus providing one of the largest collections of aeronautical and space science STI in the world. Results are published in both non-NASA channels and by NASA in the NASA STI Report Series, which includes the following report types:

- TECHNICAL PUBLICATION. Reports of completed research or a major significant phase of research that present the results of NASA programs and include extensive data or theoretical analysis. Includes compilations of significant scientific and technical data and information deemed to be of continuing reference value. NASA counterpart of peer-reviewed formal professional papers, but having less stringent limitations on manuscript length and extent of graphic presentations.

- TECHNICAL MEMORANDUM. Scientific and technical findings that are preliminary or of specialized interest, e.g., quick release reports, working papers, and bibliographies that contain minimal annotation. Does not contain extensive analysis.

- CONTRACTOR REPORT. Scientific and technical findings by NASA-sponsored contractors and grantees.

- CONFERENCE PUBLICATION. Collected papers from scientific and technical conferences, symposia, seminars, or other meetings sponsored or co-sponsored by NASA.

- SPECIAL PUBLICATION. Scientific, technical, or historical information from NASA programs, projects, and missions, often concerned with subjects having substantial public interest.

- TECHNICAL TRANSLATION. English-language translations of foreign scientific and technical material pertinent to NASA's mission.

Specialized services also include creating custom thesauri, building customized databases, and organizing and publishing research results.

For more information about the NASA STI program, see the following:

- Access the NASA STI program home page at *http://www.sti.nasa.gov*

- E-mail your question via the Internet to help@sti.nasa.gov

- Fax your question to the NASA STI Help Desk at 443-757-5803

- Phone the NASA STI Help Desk at 443-757-5802

- Write to:
  NASA STI Help Desk
  NASA Center for AeroSpace Information
  7115 Standard Drive
  Hanover, MD 21076-1320

NASA/TM-2010-216868/Volume I
NESC-RP-08-00487

Constellation Program (CxP) Crew Exploration Vehicle (CEV) Parachute Assembly System (CPAS) Independent Design Reliability Assessment

Michael J. Kelly/NESC
Langley Research Center, Hampton, Virginia

National Aeronautics and
Space Administration

Langley Research Center
Hampton, Virginia 23681-2199

November 2010

The use of trademarks or names of manufacturers in the report is for accurate reporting and does not constitute an official endorsement, either expressed or implied, of such products or manufacturers by the National Aeronautics and Space Administration.

Available from:

NASA Center for AeroSpace Information
7115 Standard Drive
Hanover, MD 21076-1320
443-757-5802

	NASA Engineering and Safety Center Technical Assessment Report	Document #: NESC-RP-08-00487	Version: 1.0
Title: CEV Parachute Assembly System Independent Reliability Analysis			Page #: 1 of 109

**Constellation Program (CxP)
Crew Exploration Vehicle (CEV)
Parachute Assembly System (CPAS)
Independent Design Reliability Assessment**

NRB Review Date: August 26, 2010

NESC Request No.: TI-08-00487

	NASA Engineering and Safety Center Technical Assessment Report	Document #: NESC-RP- 08-00487	Version: 1.0
Title:	CEV Parachute Assembly System Independent Design Reliability Assessment		Page #: 2 of 109

Report Approval and Revision History

Approval and Document Revision History

NOTE: This document was approved at the August 26, 2010, NRB. This document was submitted to the NESC Director on September 23, 2010, for configuration control.

Approved Version:	*Original Signature on File*	9/24/10
1.0	NESC Director	Date

Version	Description of Revision	Office of Primary Responsibility	Effective Date
1.0	Initial Release	Mr. Michael J. Kelly, NESC Back-Up Principal Engineer	08/26/10

NESC Request No.: TI-08-00487

NASA Engineering and Safety Center Technical Assessment Report	Document #: NESC-RP-08-00487	Version: 1.0
Title: CEV Parachute Assembly System Independent Design Reliability Assessment		Page #: 3 of 109

Table of Contents

Volume I: Technical Assessment Report

1.0 Notification and Authorization 7

2.0 Signature Page 8

3.0 Team List 9
3.1 Acknowledgements 9

4.0 Executive Summary 11
4.1 Summary of Assessment 11
4.2 Apollo Program Reports 13
4.3 Design, Development, Testing, and Evaluation (DDT&E) Guiding Principles 13

5.0 Assessment Plan 17

6.0 Problem Description, Scope of Review, and Architectural Evolution 18
6.1 Problem Description 18
6.2 Scope of Review 19
6.3 Architectural Evolution 21
    6.3.1 Review Period 1—December 2008 through March 2009 22
    6.3.2 Review Period 2—April 2009 through August 2009 37
    6.3.3 Review Period 3—September 2009 through April 2010 43
6.4 PRA and other Reliability Products 56
    6.4.1 Review Period 1—December 2008 through March 2009 56
    6.4.2 Review Period 2—April 2009 through August 2009 59
    6.4.3 Review Period 3—September 2009 through April 2010 62
6.5 Requirements, Testing, Analyses, and Verification Planning 63
    6.5.1 Review Period 1—December 2008 through March 2009 63
    6.5.2 Review Period 2—April 2009 through August 2009 63
    6.5.3 Review Period 3—September 2009 through April 2010 65
    6.5.4 Analysis 66
    6.5.5 Master Verification Plan 68
6.6 Organizational Complexity 70

7.0 Data Analysis 75
7.1 Interim Analytical Assessment of Project Information 75
    7.1.1 Review Period 1—December 2008 through March 2009 76

	7.1.2	Review Period 2—April 2009 through August 2009 77
	7.1.3	Review Period 3—September 2009 through April 2010. 78
7.2	Final Analytical Assessment of CPAS Project Information 78
	7.2.1	Architecture 79
	7.2.1.1	Pack Volume Risks 79
	7.2.1.2	Roll Control Risks 80
	7.2.2	PRA and Other Reliability Products 82
	7.2.2.1	PRA Estimation Methods 82
	7.2.2.2	Use of PRA Estimates for Design Decisions 85
	7.2.3	Requirements, Testing, Analyses, and Verification Planning 86
	7.2.3.1	Requirements 86
	7.2.3.2	Development Test Plan 87
	7.2.3.3	Ground Testing 89
	7.2.3.4	Main Pack Retention System 89
	7.2.3.5	Forward Bay Cover 91
	7.2.3.6	Parachute Near-Field Contact during Deployment 92
	7.2.3.7	Modeling and Simulation 93
	7.2.3.8	Uncertainty Analysis of Simulation Results 94
	7.2.3.9	Clustered Parachute Modeling 95
	7.2.3.10	Master Verification Plan 95
	7.2.4	Organizational Complexity Impact on Systems Integration 98

8.0	Findings, Observations, and NESC Recommendations 101
8.1	Findings 101
8.2	Observations 102
8.3	NESC Recommendations 103

| 9.0 | Lessons Learned .... 104 |

| 10.0 | Definition of Terms .... 105 |

| 11.0 | Acronyms List .... 106 |

| 12.0 | References .... 108 |

NASA Engineering and Safety Center Technical Assessment Report	Document #: NESC-RP-08-00487	Version: 1.0
Title: CEV Parachute Assembly System Independent Design Reliability Assessment		Page #: 5 of 109

List of Figures

Figure 6.1-1.	CPAS Gen-1 Con Ops	19
Figure 6.2-1.	CPAS/CEV Boundaries	20
Figure 6.3-1.	Drogue Mortars	22
Figure 6.3-2.	Drogue Harness Legs and Confluence Fitting Stowed	23
Figure 6.3-3.	Drogue Parachutes	24
Figure 6.3-4.	Main Parachute Confluence Fitting	25
Figure 6.3-5.	Main Parachute Packs Stowed	26
Figure 6.3-6.	Main Parachute Pack Retention System at IDR-1	27
Figure 6.3-7.	Main Parachutes Deploying	28
Figure 6.3-8.	Main Parachutes	29
Figure 6.3-9.	Rotation Torque Limiter Keeper for Main Harness Legs	30
Figure 6.3-10.	Rotation Torque Limiter Keeper for Main Harness Legs, Deployed	31
Figure 6.3-11.	Rotation Torque Limiter Keeper for Main Harness Legs, Deployed Detail	32
Figure 6.3-12.	Auxiliary Parachutes	33
Figure 6.3-13.	Pilot Parachutes Stowed	34
Figure 6.3-14.	Pilot Parachutes Deployed	35
Figure 6.3-15.	Drogue Harness Legs Stowed and Deployed, with Dual Confluence Rings	36
Figure 6.3-16.	Drogue Harness Leg Attachment Fittings Detail	37
Figure 6.3-17.	Zero-gusset for Main Risers Single Attach Point, First Iteration	38
Figure 6.3-18.	Zero-gusset for Main Risers Single Attach Point, Second Iteration	38
Figure 6.3-19.	Rotation Torque Limiter Keeper for Main Risers, Stowed	39
Figure 6.3-20.	Rotation Torque Limiter Keeper for Main Risers, Deployed Detail	40
Figure 6.3-21.	Main Pack Retention System at IDR-2	41
Figure 6.3-22.	Main Parachute Pack Retention System, Detail	41
Figure 6.3-23.	CM Forward Bay Arrangement at IDR-2, Detail	42
Figure 6.3-24.	Overview of IDAT Charter	43
Figure 6.3-25.	Segmented FBC and Representative Flow Analysis Image	44
Figure 6.3-26.	Segmented FBC Detail	45
Figure 6.3-27.	CM Forward Bay Detail Showing FBC Panel Circular Push-off Areas	46
Figure 6.3-28.	CM Zero-gusset Riser Flowerpot Fitting	47
Figure 6.3-29.	CM Common Gusset	48
Figure 6.3-30.	CM Backshell Angle Increase	49
Figure 6.3-31.	Main Pack Retention System	50
Figure 6.3-32.	Main Pack Retention System, Detail 1	50
Figure 6.3-33.	Main Pack Retention System, Detail 2	51
Figure 6.3-34.	Main Pack Retention System, Detail 3	51
Figure 6.3-35.	Main Pack Retention System, Detail 4	52
Figure 6.3-36.	Main Pack Retention System, Detail 5	53
Figure 6.3-37.	Main Pack Retention System, Detail 6	53

NESC Request No.: TI-08-00487

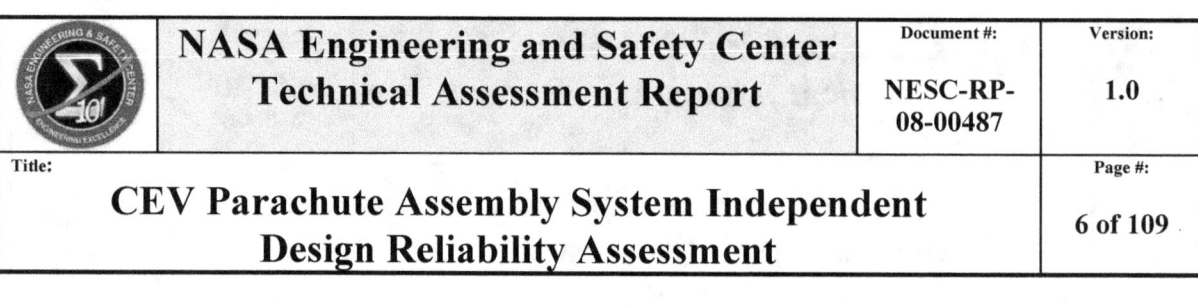

Figure 6.3-38.	Main Pack Retention System, Detail 7	54
Figure 6.3-39.	Main Pack Retention System, Detail 8	54
Figure 6.3-40.	Main Pack Ramps	55
Figure 6.5-1.	Requirements Traceability	64
Figure 6.5-2.	Representative Portion of Development Test Matrix	66
Figure 6.5-3.	Verification Strategy at IDR-3	68
Figure 6.6-1.	CPAS Organization, SE&I	71
Figure 6.6-2.	CPAS Organization, S&MA	72
Figure 6.6-3.	CPAS Project Functional Team Organization Chart	73
Figure 6.6-4.	GFE CPAS Boards and Panels Review Paths	74
Figure 6.6-5.	CPAS/ADS/LRS/CM Interface Definitions at Conclusion of RP-3	75

List of Tables

Table 6.4-1.	Drogue Parachute Reliability Estimates at IDR-1	57
Table 6.4-2.	Main Parachute Reliability Estimates at IDR-1	58
Table 6.4-3.	CPAS Input Values for PRA at IDR-2	60
Table 6.4-4.	CPAS PRA Input Value Failure Estimates at IDR-2	61
Table 6.4-5.	CPAS PRA at IDR-2	62
Table 6.5-1.	MVP Verification Strategy	69
Table 7.2-1.	Evolution of FMEA from RP-1 to RP-3	84

Volume II: Appendices (Stand-alone Volume)

Appendix A. Stakeholder Request (November 2008)

Appendix B. Stakeholder Outbrief of Interim Recommendations 1 and supporting material (April 2009)

Appendix C. Stakeholder Outbrief of Interim NESC Recommendations 2 and supporting material (September 2009)

Appendix D. Stakeholder Outbrief of Interim NESC Recommendations 3 and supporting material (April 2010)

NESC Request No.: TI-08-00487

Volume I: Technical Assessment Report

1.0 Notification and Authorization

In October 2008, Mr. Steve Altemus, Director of Engineering at the Johnson Space Center (JSC) requested an independent assessment of the Constellation Program (CxP) Crew Exploration Vehicle (CEV) Parachute Assembly System (CPAS). Mr. Altemus requested the NASA Engineering and Safety Center (NESC) create a cross-Agency and multidiscipline team to examine the following:

- Review the current CPAS's design, reliability estimates, and development test plans
- Identify design improvement opportunities to increase the overall CPAS reliability
- Identify improvements to the development test plan (revisions, additions, or deletions) that will improve its efficiency and effectiveness
- Identify process improvement opportunities to increase the overall CPAS reliability

An NESC out-of-board activity was approved by Mr. Ralph Roe, the NESC Director, on October 22, 2008. Mr. Michael Kelly, NESC Back-Up Principal Engineer at Langley Research Center (LaRC), was selected to lead this assessment. The assessment plan was approved by the NESC Review Board (NRB) on November 20, 2008. The assessment plan was subsequently updated and approved by the NRB on March 26, 2009, and August 20, 2009.

The key stakeholders for this assessment were the CxP CPAS Systems Engineering and Integration (SE&I) lead, Mr. Timothy Fisher; the CPAS Chief Engineer, Mr. Ricardo "Koki" Machin; and the CPAS Safety and Mission Assurance (S&MA) lead, Ms. Christine Stewart.

2.0 Signature Page

Submitted by:

Team Signature Page on File – 11/12/10
_____
Mr. Michael J. Kelly   Date

Significant Contributors:

_____   _____
Dr. Vance L. Behr   Date   Mr. Douglas C. Brown   Date

_____   _____
Mr. John E. Hengel   Date   Mr. Michael Herr   Date

_____   _____
Mr. Jerry E. McCullough   Date   Dr. Peter A. Parker   Date

_____   _____
Mr. Jeremy D. Shidner   Date   Mr. Robert B. West   Date

_____   _____
Dr. James M. Womack   Date   Mr. Gary K. Won   Date

Signatories declare the findings and observations compiled in the report are factually based from data extracted from Program/Project documents, contractor reports, and open literature, and/or generated from independently conducted tests, analysis, and inspections.

3.0 Team List

Name	Discipline	Organization/Location
Civil Servants		
Michael Kelly	Assessment Lead	LaRC
Juan Cruz*	Parachute Systems DDT&E	LaRC
John Hengel	Parachute Systems DDT&E	MSFC
Frederick Huegel	Electrical and Avionics	GSFC
Natesan Jambulingam	Reliability and Statistics	GRC
Brian Jensen	Materials and Process	LaRC
Peter Parker	Statistics and DOE	LaRC
Karma Snyder	Test Engineering and Systems Engineering	SSC
Walter Thomas	Reliability and Statistics	GSFC
Tricia Johnson	MTSO Program Analyst	LaRC
Contractors		
Vance Behr	Parachute Systems DDT&E	Sandia National Laboratories
Douglas Brown*	Reliability and DOE	Booz Allen Hamilton
Michael Herr	Parachute Systems DDT&E	NAWS-China Lake
Jerry McCullough*	Systems Safety and Test Engineering	SAIC
Jeremy Shidner	Modeling and Simulation	AMA/LaRC
Nicholas Vitullo*	Structures Engineering and Systems Engineering	ATK/LaRC
Robert West	Parachute Systems DDT&E	SAIC
James Womack	Reliability and Statistics	The Aerospace Corporation
Gary Won	Electrical, Avionics, Systems Engineering	BEI/GSFC
Administrative Support		
Linda Burgess	Planning and Control Analyst	ATK/LaRC
Tina Dunn-Pittman	Project Coordinator	ATK/LaRC
Carolyn Snare	Technical Writer	ATK/LaRC

Team members participated for less than the full term of the assessment.

3.1 Acknowledgements

The NESC assessment team would like to acknowledge and thank members of the entire CPAS organization including NASA, Jacobs Engineering and Science Contract Group (Jacobs ESCG),

Airborne Systems (AS), and Lockheed Martin (LM), for their tolerance of assessment team members quietly 'lurking' on so many of their teleconferenced meetings.

Thanks are given to Mr. Timothy Fisher and Mr. Michael Hazen, the CPAS SE&I co-leads from NASA and Jacobs ESCG. The assessment team also thanks Ms. Christine Stewart and Mr. Achilles Soukis from the CPAS S&MA organization. Special gratitude is owed to Ms. Charlyn Josey, NASA SE&I administrative assistant, for her exceptional responsiveness to NESC assessment team requests for CPAS information.

The team also acknowledges the outstanding support from Alliant Techsystems, Inc. (ATK)'s Project Coordinator Mrs. Tina Dunn-Pittman, who kept up with voluminous amounts of information discussed during team meetings. ATK Planning and Control Analyst Ms. Linda Burgess easily managed two schedule changes and an assessment hiatus. Special thanks are given to ATK Technical Writer Ms. Carolyn Snare for her tremendous help getting three *interim* stakeholder outbriefs and one *final* stakeholder outbrief completed on time, and not least for getting this two-volume final report completed and with excellent appearance.

The considerable effort and quality feedback provided by three peer reviewers, Mr. Steven Gentz, Mr. Daniel Yuchnovicz, and Mr. Kenneth Johnson, are also acknowledged with tremendous gratitude.

The assessment team leader wishes to thank the entire assessment team for the amazing education he received in parachute design development and reliability methodology.

4.0 Executive Summary

4.1 Summary of Assessment

This report documents the activities, findings, and NASA Engineering and Safety Center (NESC) recommendations of a multidiscipline team to independently assess the Constellation Program (CxP) Crew Exploration Vehicle (CEV) Parachute Assembly System (CPAS). This assessment occurred during a period of 15 noncontiguous months between December 2008 and April 2010, prior to the CPAS Project's Preliminary Design Review (PDR) in August 2010.

The assessment team's planned objectives were to identify reliability-enhancing opportunities for the CPAS Project:

- Parachute system design
- Reliability estimation methodologies
- Development program (development tests and analysis, and verification plans)
- Design development and integration practices

The assessment team's activities were divided into three review periods (RPs). Each RP began with an Internal Design Review (IDR) and concluded with a stakeholder summary (originally titled stakeholder outbrief) of findings, observations, and *interim* NESC recommendations.

RP-1   December 2008–March 2009
RP-2   April 2009–May 2009 and July–August 2009
RP-3   September 2009–April 2010

Prior to 2008, the CPAS architecture (referred to as the "Generation-1" (Gen-1) architecture) included two mortar-deployed drogue parachutes and three mortar-deployed pilot parachutes; each pilot parachute deployed a main parachute. The concept of operations (Con Ops) had the crew module's (CM's) forward bay cover (FBC) jettisoning as the first event, prior to firing the drogue mortars.

During 2008, the CPAS architecture was modified such that the initiating event was the mortar-deployment of the two drogue parachutes through holes in the FBC. The drogues were to share a common, three-legged harness that was stowed externally on top of the FBC. Subsequent jettison of the FBC would deploy three main parachutes attached to its underside by virtue of the separation event. This was the CPAS's architecture when the NESC team began its assessment.

During the assessment, the Project considered, studied, adopted, and discarded numerous architectural options. These are described in detail in Section 6.3 and assessed in Section 7.2.1. By the end of the assessment, the final "Generation-2" (Gen-2) architecture had been selected and work was proceeding toward the Project's PDR in August 2010. This architecture resembled the Apollo Program Earth Landing System (ELS), with two mortar-deployed drogue parachutes, three mortar-deployed pilot parachutes, and three pilot-parachute-deployed main parachutes. Drogue and main parachute risers were attached to the CM at a common location and were steel rather than textile.

During the assessment, the Project's fault tree analysis (FTA) and probabilistic risk assessment (PRA) also changed considerably, commensurate with the architecture changes (as detailed in Section 6.4 and assessed in Section 7.2.2). The requirements and development plans also evolved with the architecture (as detailed in Section 6.5 and assessed in Section 7.2.3).

The NESC assessment team outbriefed findings, observations, and *interim* NESC recommendations to the Project at the end of each RP. In all, 55 *interim* NESC recommendations were provided. Most of these were acted upon in part or in whole by the Project during the assessment period or were overcome by ongoing Project design decisions. Some *interim* NESC recommendations from the third RP are reiterated as final recommendations in this report. (See Appendices B, C, and D for a complete list of *interim* NESC recommendations.)

This final report provides findings, observations, and nine *final* NESC recommendations that pertain to the Gen-2 CPAS architecture. All are reiterations or enhanced restatements of previously approved *interim* NESC recommendations that were of the top-most concern to the assessment team at the conclusion of its assessment.

Two final NESC recommendations pertain to the architecture itself. The assessment team reiterates an interim recommendation that the Project should establish a formal volume budget for managing parachute growth throughout the development effort and even after Orion becomes operational. The assessment team also recommends that the Orion Project comprehensively reevaluate the many integrated issues that arise from the reliance on a CM roll reorientation maneuver just prior to water landings. Both of these issues may impact CPAS reliability.

The assessment team makes two final NESC recommendations pertaining to the PRA. One recommends the preferential use of certain data over others for probability failure estimates, and the other restates an interim recommendation about conveyance of data sources and uncertainties when PRA results are communicated to decision makers.

The assessment team presents one final NESC recommendation on requirements planning and restates another recommendation regarding the utility of statistical design-of-experiment (DOE) techniques for planning an effective and efficient development program.

The assessment team collected several findings about risk associated with specific design details in the Gen-2 architecture. They combined them in one final NESC recommendation for specific near-term ground tests to provide information that will enable the Project to make more informed design decisions.

One other interim recommendation is reiterated regarding the conveyance of the validity and uncertainty associated with analytical results when they are communicated to decision makers.

Finally, an NESC recommendation is made for the application of DOE to a comprehensively planned verification effort to assure CPAS matures as a robust, reliable system.

4.2 Apollo Program Reports

A significant recurrent observation of the assessment team was that the Project's PRA lacked sufficient appropriate data on which to base failure probability estimates (see Section 7.2.2 for more details). The assessment team found that a key early estimate was inappropriately based on Space Shuttle Program (SSP) Solid Rocket Booster (SRB) service data. Later, the Project applied data from an expert elicitation (EE) exercise with which the assessment team found shortcomings. With the final Gen-2 architecture now resembling the Apollo Program ELS in many ways, the assessment team concluded that applicable data from Apollo tests would be of tremendous utility. Unfortunately, many of the contractor reports generated during ELS development could not be located within NASA. This revelation led the assessment team to create an Agency lesson learned that NASA Programs and Projects should not rely on contractor reports surviving but rather should document development efforts themselves and store the documentation in a topic-specific or discipline-specific Agency repository.

During the assessment period, a cache of parachute testing and reliability reports from the now-defunct Northrop Ventura company was discovered in a Northrop Grumman (NG) document storage facility. A contract with NG was funded with assessment budget to allow a retired (Rockwell) expert on heritage Apollo parachute testing (not an assessment team member), who had first-hand knowledge of these reports, to conduct two surveys to inventory these reports at the storage facility in August and December 2009. The assessment team's business analyst also facilitated a subsequent contract (funded by the Orion Project) to acquire digitized copies of critical reports identified in the survey. This contract was in place in July 2010. The assessment team considered the data in these reports essential to creating a valid CPAS PRA and helping ensure an efficient development test program. These digitized reports will be stored in perpetuity in an NESC-sponsored, Agency-wide repository of entry, descent, and landing (EDL) data.

4.3 Design, Development, Testing, and Evaluation (DDT&E) Guiding Principles

During the assessment the assessment team discussed design development best practices.

The assessment team provided a reference copy of NASA-TM- 2008-215126, *Design, Development, Testing and Evaluation (DDT&E) - Considerations for Safe and Reliable Human Rated Spacecraft Systems* [Ref. 1] to the Project at its first interim outbrief in April 2009. The report provides seven guiding principles that apply to those who conceive, produce, and operate integrated systems, individual components, or system elements. The guiding principles are supported by a foundation of established project management, systems engineering, S&MA, and operations practices:

1. Define a clear and simple set of prioritized program needs, objectives and constraints, including safety, that form the validation basis for subsequent work.
2. Manage and lead the program with a safety focus, simple and easy to understand management structures, and clear lines of authority and responsibility among the elements.
3. Specify safety and reliability requirements through a triad of fault tolerance, bounding failure probability, and adhering to proven practices and standards.
4. Manage complexity by keeping the primary (mission) objectives as simple and minimal as possible and adding complexity to the system only where necessary to achieve these objectives.
5. Conceive the right system conceptual design early in the life cycle by thoroughly exploring risks from the top down and using a risk-based design loop to iterate the operations concept, the design, and the requirements until the system meets mission objectives at minimum complexity and is achievable within constraints.
6. Build the system right by applying a multilayered, "defense in depth" approach of following proven design and manufacturing practices, holding independent reviews, inspecting the end product, and employing a "test like you fly, fly like you test" philosophy.
7. Seek and collect warning signs and precursors to safety, mission success, and development risks throughout the life cycle and integrate those into a total risk picture with appropriate mitigation activities.

The assessment team observed numerous examples of the Project implementing the guiding principles recommended in the DDT&E report.

During the assessment the Project exhaustively examined top-level architecture options, explored risks from the top down, and employed a risk-based design loop to iterate the operations concept, the design, and the requirements until the system appeared to meet mission objectives at minimum complexity and within achievable constraints. The assessment team observed numerous positive examples of curiosity, skepticism, and imaginative inquiry during several CPAS Project meetings. These practices were consistent with elements of DDT&E guiding principle 2.

The assessment team also observed commendable Project application of DDT&E guiding principal 3. The DDT&E report notes that Projects should consider off-nominal and failure scenarios to generate requirements that seek to prevent faults and ensure crew safety in spite of faults. The report advises that to maximize safety and reliability, systems should be designed with a fault-tolerant architecture that is supported by probabilistic safety, reliability, and risk analyses supported by data, evidence, and/or analysis data. The assessment team observed this in Project practice, although it noted occasions of application of un-validated analysis results in the process.

The DDT&E report also notes (in supportive narrative for guiding principal 3) that the objective of bounding a system's probability of failure during development is to encourage a thorough investigation into risks including uncertainty and common cause such that system design decisions and underlying risk analysis can be defended. The report further notes that the process of estimating failure probabilities should challenge design teams to achieve deeper understanding of its system and its environment and facilitate design trades while protecting against worst-case assumptions that can cause a design to become overly complex. The DDT&E report advises

> "The value of a probability estimate is not so much contained in the absolute number but in the thorough investigation, debate, and discussions by designers and operators about controlling the potential for failures based on their likelihood, history of similar systems, and uncertainties inherent in the system design. The analysis of the system design must consider the integrated whole and include a top-down assessment. The analyses are most useful for evaluating and comparing design and operations alternatives and validating the chosen system design. To ensure valid comparison correct statistical methods should be used to determine probability of failure and include all available data sources."

The assessment team observed exemplary application of this practice by the Project, although it noted instances of improper use of statistical methods in the process.

Guiding principal 4 in the DDT&E report encourages Projects to keep primary mission objectives as simple and minimal as possible and to add complexity to a system only where necessary to achieve those objectives. The assessment team observed with satisfaction that the Gen-2 architecture is less complex than several that were considered.

However, the assessment team observed that the complexity of the CPAS management structures and its lines of authority and responsibility were sub-optimal and that the observed complexity was contrary to elements of DDT&E guiding principal 2.

CPAS was government-furnished equipment (GFE) with Jacobs ESCG contracted to manage AS as the Project's design developer. LM was responsible for parachute mortars, riser cutters, and other elements the assessment team thought were critical to CPAS reliability. This is described in more detail in Section 6.6 and the team's assessment can be found in Section 7.2.4.

The assessment team felt strongly that fragmenting DDT&E responsibilities among multiple entities posed significant risks to CPAS reliability. The assessment team thought Orion should have empowered a single entity with the responsibility for the design, development, integration, and certification (DDI&C) of the entire Landing Recovery System (LRS). It was recognized that this multi-element organization was unlikely to change for Orion. The risks associated with organizational complexity were generally recognized by the Project and systems engineering practices were being employed in an effort to manage those risks.

The assessment team created an Agency lesson learned for future Programs that includes development of parachute architectures for recovery of human-rated space vehicles. The lesson would apply whether the system was developed by NASA or by a commercial company with NASA oversight: Developers of parachute architectures for recovery of human-rated space vehicles should empower a single entity with the responsibility for the DDI&C of the entire LRS. The responsibility should include primary and secondary structures, control systems, crew protection systems, parachutes, mortars, and any impact-attenuation systems.

The assessment team concludes that the CPAS Project can deliver a robust and reliable CPAS with the Gen-2 architecture and develop it in a cost-effective manner, if the Project rigorously adheres to DDT&E best practices and guiding principles and if the NESC recommendations in Section 8.3 of this report are considered.

5.0 Assessment Plan

In November 2008, the Director of Engineering for JSC requested the NESC form a cross-Agency and multidiscipline team to independently assess the CPAS design, the CPAS Project's PRAs, and the methods used to derive them. The assessment team was also asked to assess the Project's development test plans and design development practices that might affect the system reliability. The requestor asked for recommendations that could "strike the proper balance between reliability enhancing (development) tests and… project cost and schedule." Cost and schedule analyses were out of scope for this assessment, but the assessment team did analyze means to improve the *efficiency* of the Project's test plans. See Appendix A for the stakeholder request.

The assessment's objectives were to identify reliability-enhancing opportunities for the:

- Parachute system design
- Reliability estimation methodologies
- Development program (development tests and analysis, and verification plans)
- Design development and integration practices

The assessment team consisted of seven contract consultants, eight NASA civil servants from five NASA Centers, and two team members from other United States (U.S.) Government entities. The core team had two subteams: one for parachute design development and one for reliability. Some team members had expertise in disciplines or systems pertinent to the assessment including materials, modeling and simulation, structures, avionics, and electrical. Other team members contributed expertise in systems integration, systems safety, test engineering, and experimental design. All assessment team members worked part time, generally less than 10 hours per week.

The overall assessment time was 15 noncontiguous months between December 2008 and April 2010. The team's assessment activities were divided into three RPs:

- RP-1   December 2008–March 2009
- RP-2   April 2009–May 2009 and July–August 2009
- RP-3   September 2009–April 2010

RP-2 included a planned assessment hiatus from May–July 2009. This stand-down was due to slippages in the Project schedule. The hiatus allowed the Project to make progress toward a more concrete design architecture, corresponding reliability work products, and a draft development test plan. The stand-down also allowed the preservation of assessment budget.

During this assessment, the team monitored Project meetings and reviewed Project presentation charts, draft documents, test reports, and other technical material. The assessment team attended three Project IDRs [Refs. 2, 3, and 4], one at the beginning of each RP. The assessment team conducted weekly discussions to identify findings consistent with the plan objectives.

Parachute development subteam members and reliability subteam members actively taught each other aspects of their respective crafts. Assessment team members with experience on the Apollo Program, SSP SRB, and military parachute systems shared program-specific experiences that were pertinent to CPAS discussions. Short tutorials were provided by some team members for the benefit of others:

- Introduction to parachutes (Cruz)
- Parachute definitions, nomenclature, and types (Behr)
- Reefing line cutters (Herr)
- Design of Experiments (Parker)

Assessment team members not on the parachute subteam also referred to the *Parachute Recovery Systems Design Manual*, T. W. Knacke (1992) [Ref. 5] to understand parachute development terms and conventions. Multiple copies of this 'parachute bible' were made available by team member Michael Herr (whose organization, the U. S. Navy, owns the rights to publish this document).

The assessment team outbriefed findings, observations, and *interim* NESC recommendations to the Project at the end of each RP and provided written stakeholder summaries of these to the requester. In all, 55 *interim* NESC recommendations were provided. Most of these recommendations were acted upon in part or in whole by the Project over the course of the assessment, or they were overcome by ongoing Project design decisions. Clarifying material for two interim NESC recommendations related to experimental design techniques was also provided to the Project. This material included questions that the assessment team had heard asked by Project members during Project meetings, with answers from the assessment team. The three stakeholder summaries and associated material are located in Appendices B, C, and D.

This final report provides findings, observations, and *final* NESC recommendations that pertain to the Gen-2 CPAS architecture, which was current at the conclusion of the assessment period in April 2010.

6.0 Problem Description, Scope of Review, and Architectural Evolution

6.1 Problem Description

The original CPAS architecture was derived and scaled up from the Apollo Program ELS. The design included two mortar-deployed drogue parachutes, three mortar-deployed pilot parachutes, and three pilot-parachute-deployed main parachutes. NASA and AS, its subcontracted design development company, conducted path-finding development drop testing of this CPAS system, known as Gen-1, prior to calendar year 2008 (CY08) (see Figure 6.1-1).

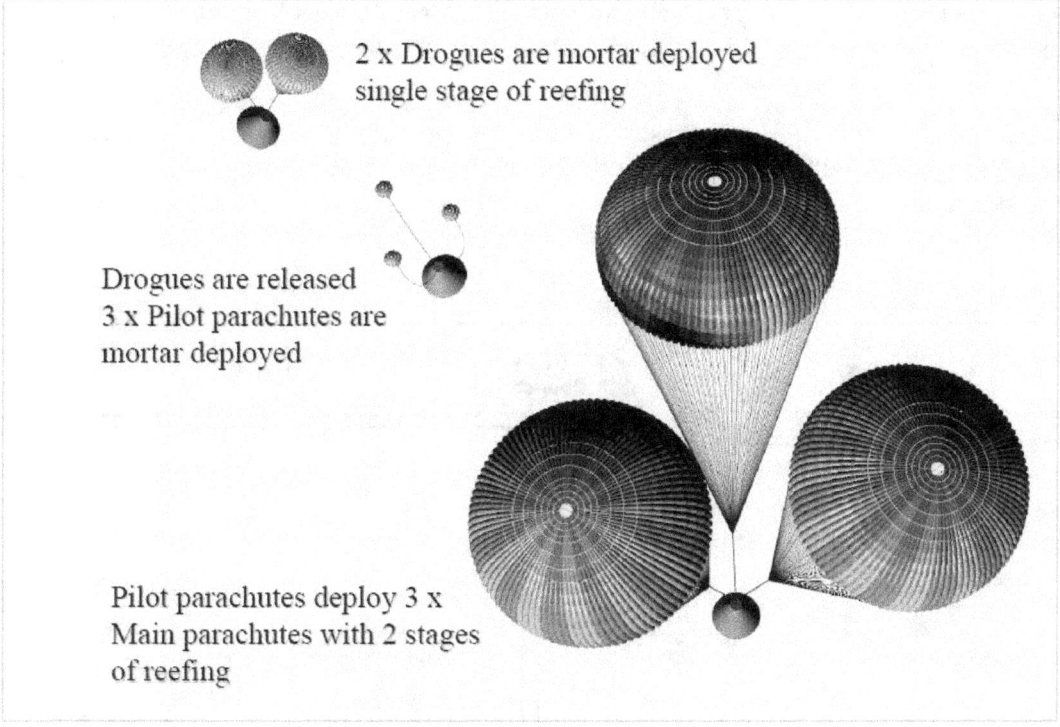

Figure 6.1-1. CPAS Gen-1 Con Ops

In early 2008, Project managers changed the CPAS architecture to what was then reasoned to be a simpler and more reliable design. This architecture included two mortar-deployed drogue parachutes and three main parachutes. The main parachutes were physically deployed by the separation of Orion's FBC from the vehicle.

In September 2008, the Orion Engineering Review Board (ERB) identified the CPAS design as the top risk contributor for loss of crew (LOC) for the Orion vehicle, based on a CPAS Project estimate that the probability of catastrophic damage to more than one main parachute during deployment was 1:217. This ERB finding precipitated the request for this NESC independent assessment.

6.2   Scope of Review

The technical assessment's planned objectives are described in Section 5.0. Taking a systems engineering approach, the scope of the NESC assessment team's hardware purview extended beyond CPAS components, to include components of other systems that were considered critical to the reliable function of CPAS. These included mortars, riser cutters, and avionics. Certain Orion structural elements and features were also in-scope, including its FBC, forward bay gussets, docking tunnel, and riser attach hardware. NASA Standard Initiator (NSI) reliability was also reviewed. Most remaining Orion systems and structures were beyond the scope of the assessment team's review.

Figure 6.2-1 shows an early Project assessment of the general boundary between CPAS and the Orion CEV, as identified by the Project at the beginning of RP-1. The figure is illustrative of issues to be identified later in this report but is not all inclusive. It does not for example include deployment bags, retention systems, or other CPAS components.

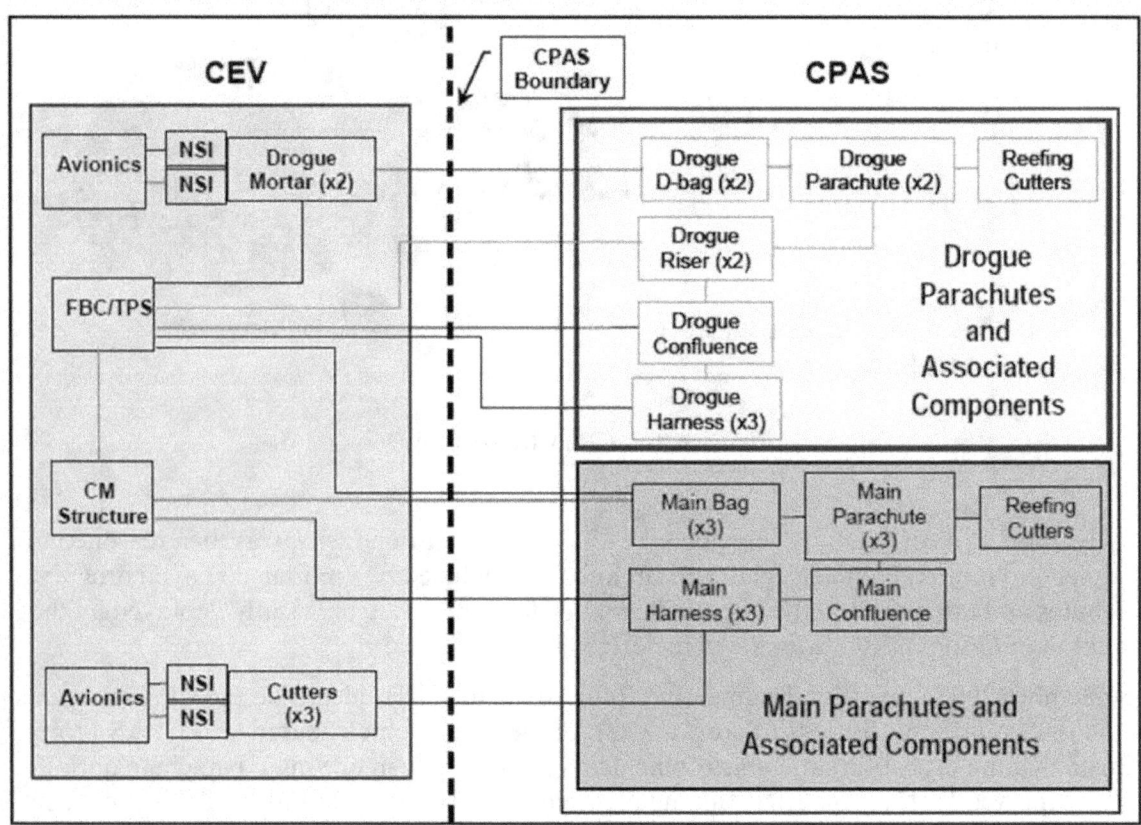

Figure 6.2-1. CPAS/CEV Boundaries

The scope of the assessment team's review also included Project reliability work products, principally but not limited to, FTAs and PRAs. The Project's development test plans and planned verification activities were also in-scope. The assessment team also considered aspects of organizational structure, and Project design development and integration practices to be within the scope of its assessment, inasmuch as they may affect reliability.

Section 6.0 details the chronological evolution of the CPAS architecture, the Project's PRA and other reliability products, and the Project's requirements, testing, analyses, and verification plans. The section is organized by the three RPs that encompassed the entire period of assessment. These are aligned with three interim stakeholder summaries which can be found in Appendices B, C, and D. A subsection is also included that describes the Project's organizational structure.

Section 7.0 contains the assessment team's analysis of the information in Section 6.0 as it pertains to the *final* findings, observations, and NESC recommendations listed in Section 8.0.

6.3 Architectural Evolution

The CPAS design evolved substantially during the overall assessment period, but the assessment team's objectives and scope remained unchanged.

The assessment team spent time early in the assessment reviewing the architectures of heritage Apollo Program ELS, SSP SRB, and CxP Ares I recovery systems. The assessment team also discussed aspects of some planetary spacecraft parachute systems. Parachute design development subteam members gave tutorials on mortar and reefing cutter designs and function for the benefit of other team members less knowledgeable of parachute systems. The assessment team also discussed parachute canopy materials applications and material joint testing.

During early team meetings, the assessment team discussed CPAS interfaces with the CM. The assessment team also discussed information in the *Apollo Experience Report, Earth Landing System (ELS)*, NASA TN D-7437 (1973) [Ref. 6], written by the heritage Apollo parachute test manager on the assessment team. The assessment team discussed at length the differences between the CPAS and ELS drogue deployment envelopes, system layouts, and components. The assessment team also reviewed the Orion Launch Abort System (LAS) design and its separation Con Ops as it impacted CPAS reliability following an abort.

The CPAS architecture evolution during the three assessment RPs are described chronologically in this subsection.

6.3.1 Review Period 1—December 2008 through March 2009

RP-1 spanned December 2008 through March 2009. The Project conducted IDR-1 in December 2008. At that time, the CPAS PDR was scheduled for June 2009. The CPAS architecture and nominal concept of operations were as follows:

During descent through the atmosphere after reentry, or following a pad abort or an ascent abort, the initiating event in the CPAS Con Ops was the simultaneously electrical triggering of two drogue mortars. The mortars were located in adjacent sectors among the six sectors of the CM forward bay and were aimed outward approximately 45 degrees apart from each other and approximately normal to the CM conic outer mold line (OML). The mortars were physically mounted to holes in the CM FBC and fired their drogue parachutes through those holes (see Figure 6.3-1).

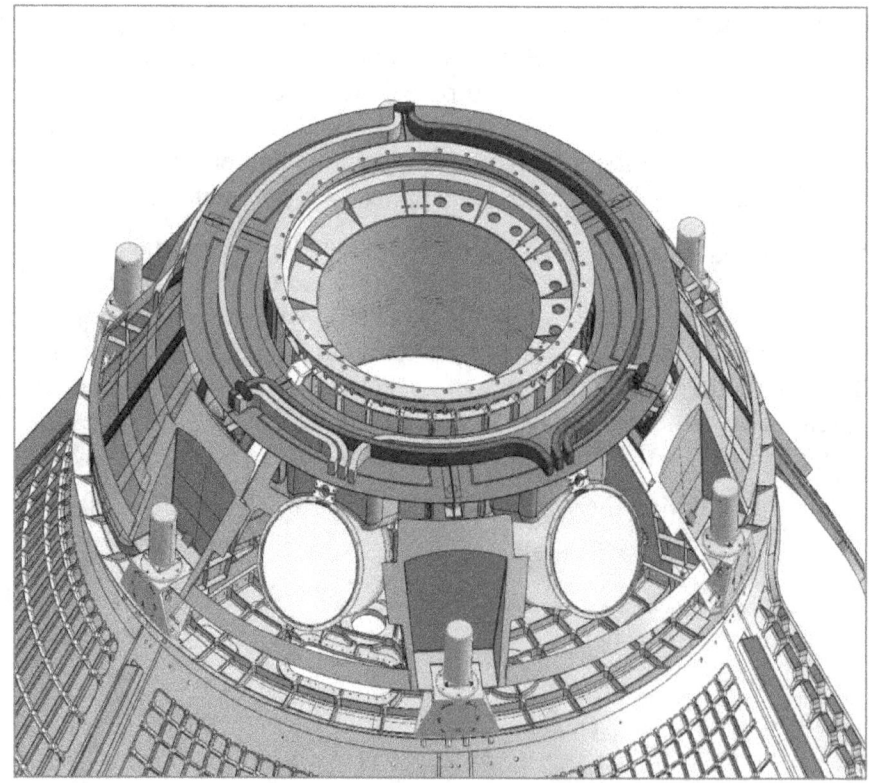

Figure 6.3-1. Drogue Mortars

Each mortar ejected a single drogue parachute and a length of textile riser into the free stream flow through which the CM was descending, at nominally 120 f/s. The riser length was the first item out of a mortar, followed by the packed drogue parachute and then the mortar sabot. Both risers were attached to a single drogue confluence fitting. Attached to the drogue confluence fitting, in addition to the two drogue risers, were three textile harness legs. The opposite end of each drogue harness leg was attached to a fitting on the CM FBC. The three attach fittings were 120 degrees apart. Drogue riser loads were reacted to the CM through the FBC. The three harness legs and the drogue confluence fitting were stowed in recessed circumferential cavities on top of the FBC, along with remaining lengths of drogue riser, covered with thermal protection system (TPS) material (see Figure 6.3-2).

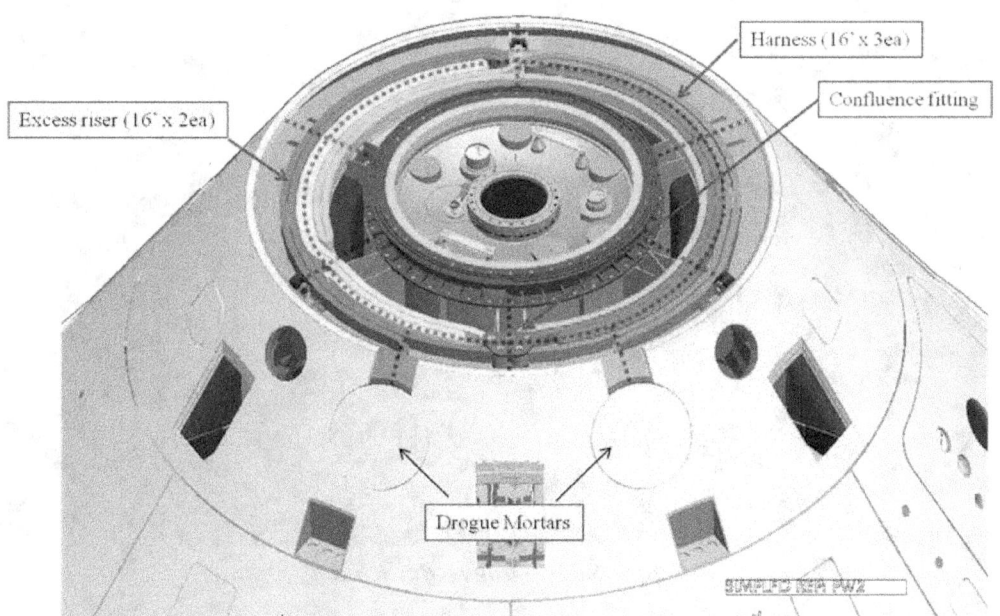

Figure 6.3-2. Drogue Harness Legs and Confluence Fitting Stowed

In the Con Ops for this architecture, each mortar-ejected drogue parachute would begin to open as it reached 'line stretch,' when its riser and suspension lines are fully deployed (see Figure 6.3-3). After this point, the drogue parachutes would dis-reef twice to their full-open position, using redundant reefing line cutters. The drogue parachute canopies were conical canopy ribbon parachutes with a nominal diameter of 23 ft. They had a continuous 2.0-in.-wide ribbon with one splice per ribbon and no more than two splices on any one gore.

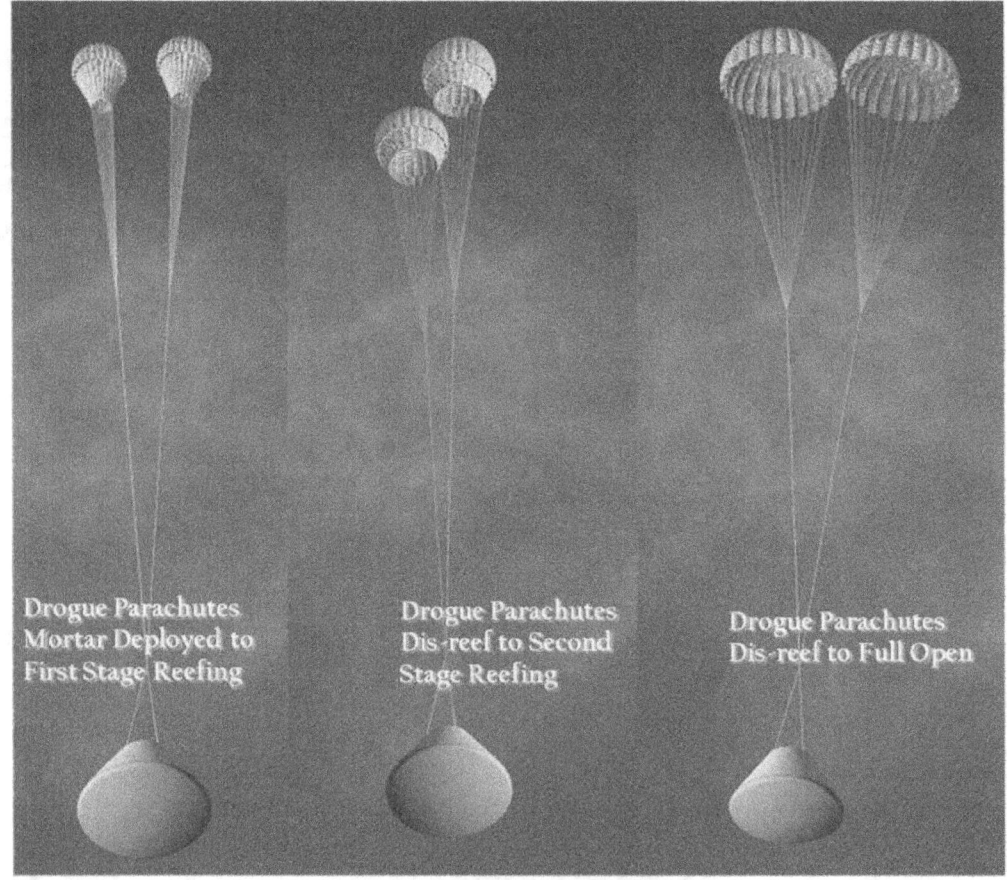

Figure 6.3-3. Drogue Parachutes

At a preprogrammed, sensed altitude, the FBC would be mechanically released from the CM and the CM would fall away from the drogue-suspended FBC. In this way the drogue parachutes provided the forces required to separate the FBC from the CM.

Each of the three main parachutes was packed in its own protective deployment bag (d-bag). The three main d-bags were restrained with textile straps to the CM upper deck, one in each of three of the six sectors of the CM forward bay. The stowed arrangement of the three d-bags was asymmetric. The upper surface of the main d-bags was loosely attached by textile straps to the underside of the FBC, at points beneath the drogue attach points. Energy-modulating textile devices were incorporated into these straps to dissipate snatch loads that would result when the CM rapidly fell away from the decelerated FBC.

In this architecture, the main parachute suspension lines and portions of their risers were also stowed inside the main d-bags. The bottom ends of the three main risers were attached to a single main parachute confluence fitting (see Figure 6.3-4) that was stowed in a stowage container in one sector of the CM forward bay.

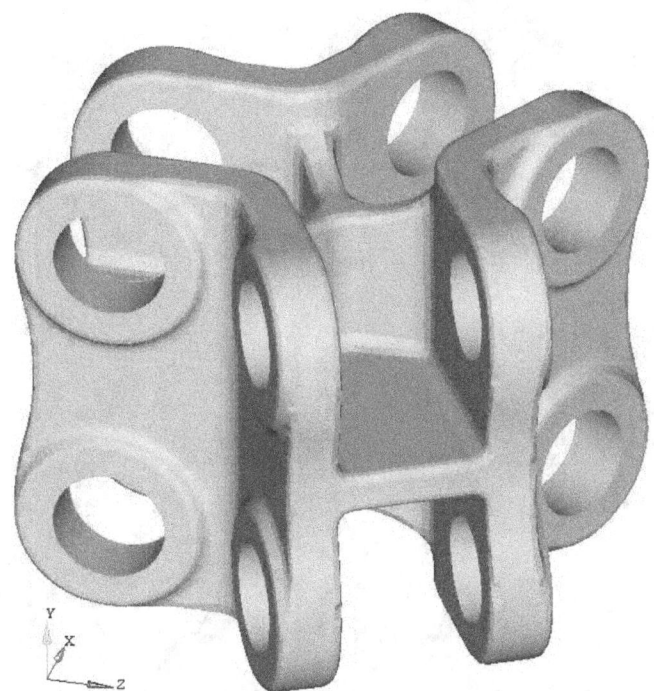

Figure 6.3-4. Main Parachute Confluence Fitting

Three textile harness legs were also attached to the main confluence fitting. One harness leg was shorter than the other two, to provide a desired hang angle for limiting CM loads during water landings. Most of the lengths of the harness legs were to be stowed in the same stowage container as the confluence fitting (see Figure 6.3-5). The bottom end of each main harness leg was attached to a fitting on the CM. The three CM attach fittings were 120 degrees apart, located atop three of the six forward bay gussets. These locations were clocked 60 degrees from the locations of the three FBC fittings.

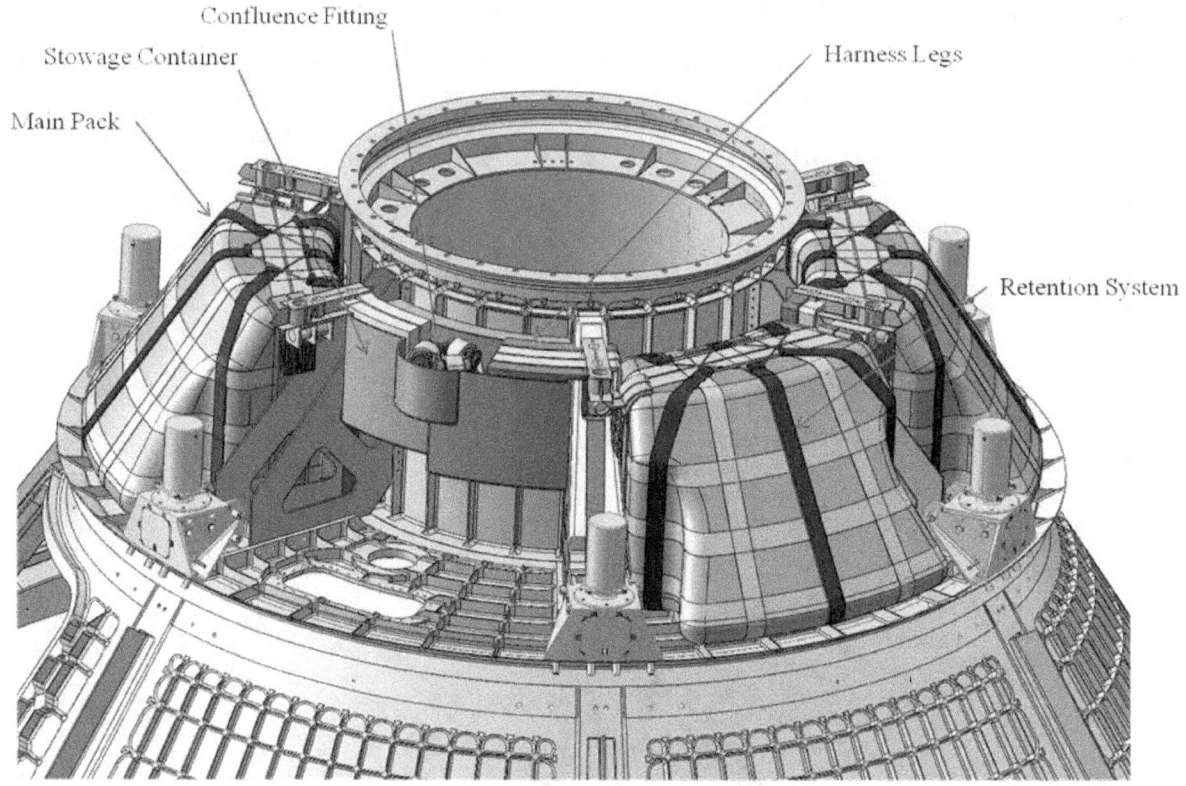

Figure 6.3-5. Main Parachute Packs Stowed

As the CM was released from the monolithic FBC, two redundant donut ties attached at the ends of six Kevlar® retention straps were to be cut by redundant, electrically commanded cutters (see Figure 6.3-6).

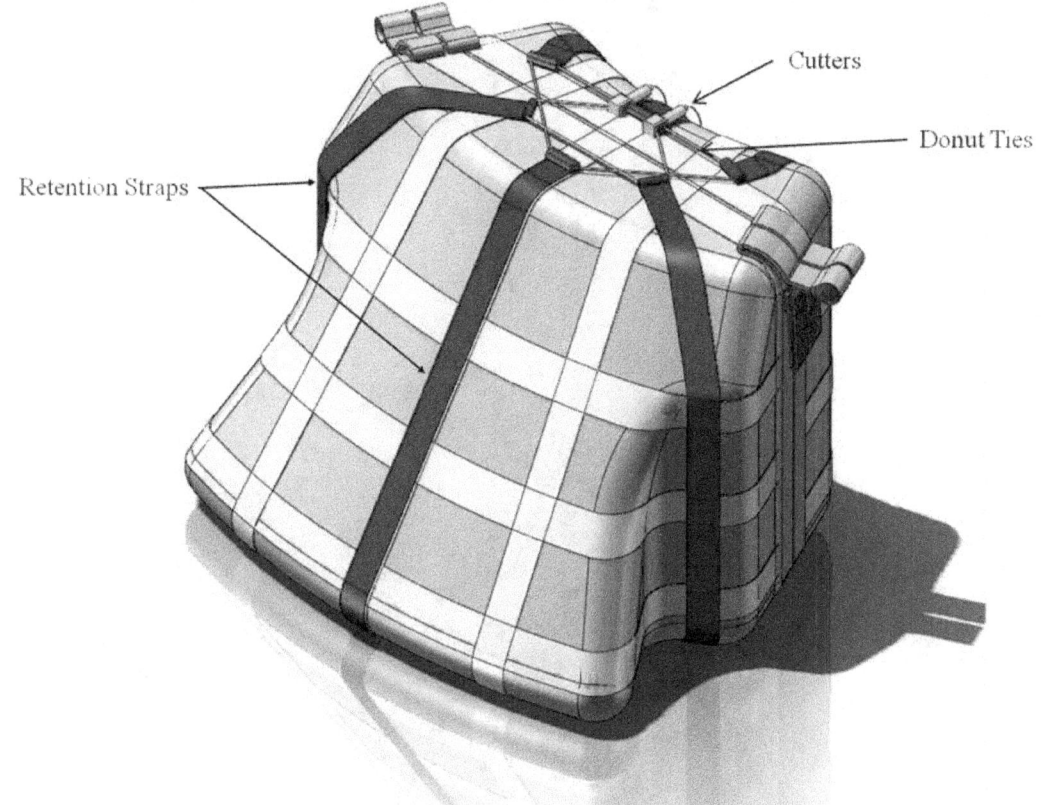

Figure 6.3-6. Main Parachute Pack Retention System at IDR-1

The separation action of the CM free-falling away from the drogue-suspended FBC would then physically lift the three main d-bags from their stowed locations by two textile handles. These handles had textile energy modulators incorporated to reduce the loads associated with what would otherwise be large pack acceleration at the moment of extraction.

As the FBC/CM separation increased, the main risers, suspension lines, and parachute canopies were sequentially pulled from inside their respective d-bags. Simultaneous to this, the main confluence fitting was lifted from its stowage container to its deployed position (see Figure 6.3-7).

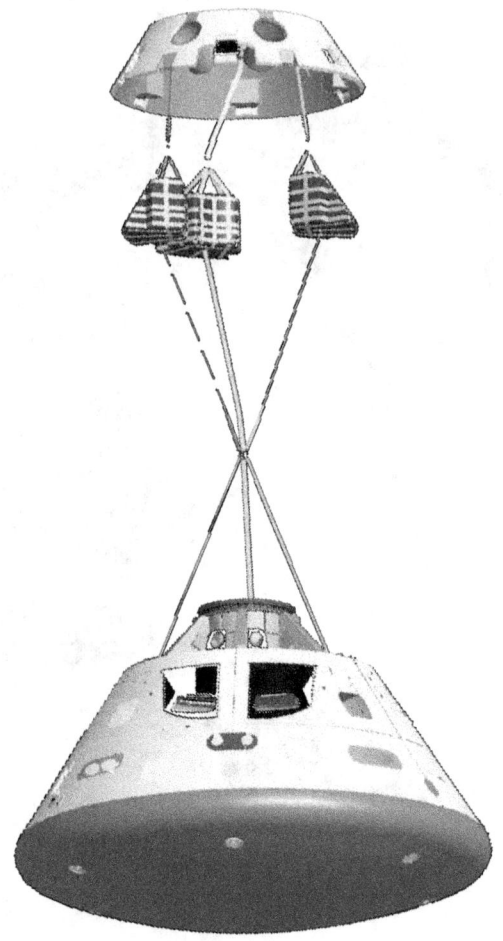

Figure 6.3-7. Main Parachutes Deploying

Ultimately the three empty main d-bags were lifted away with the FBC, and the three fully deployed main parachutes would begin to inflate.

At this point, the main parachutes would dis-reef twice to their full-open position, using redundant reefing line cutters. The main parachute canopies were ringsail parachutes with a nominal diameter of 116 ft (see Figure 6.3-8).

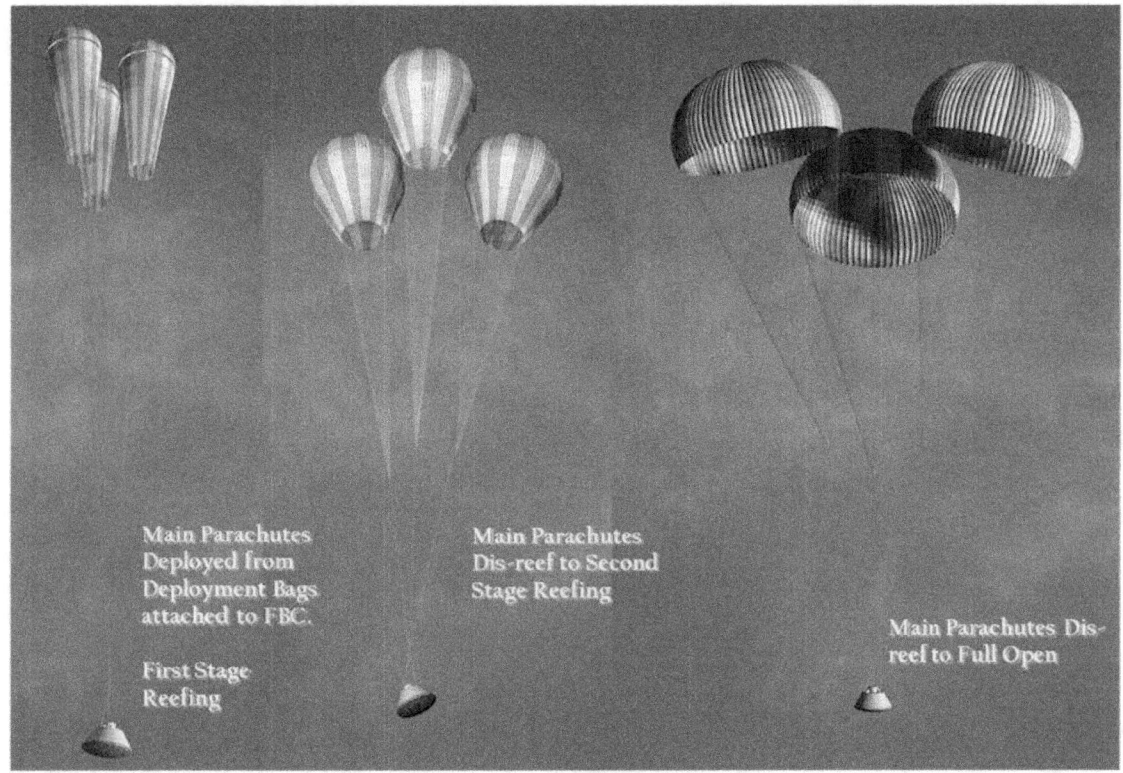

Figure 6.3-8. Main Parachutes

One other key feature installed along the main parachute risers (not shown in previous figures) was the rotation torque limiter (i.e., keeper) (see Figure 6.3-9). This triangular metallic device was included to limit roll torque resistance to below 450 ft·lbf when the CM was suspended under its opened main parachutes. This feature provided a means for the CM reaction control system (RCS) to have sufficient authority to roll the CM into a toe-forward position just prior to water landing. This landing maneuver was a method to reduce water-landing design loads on the CM structure.

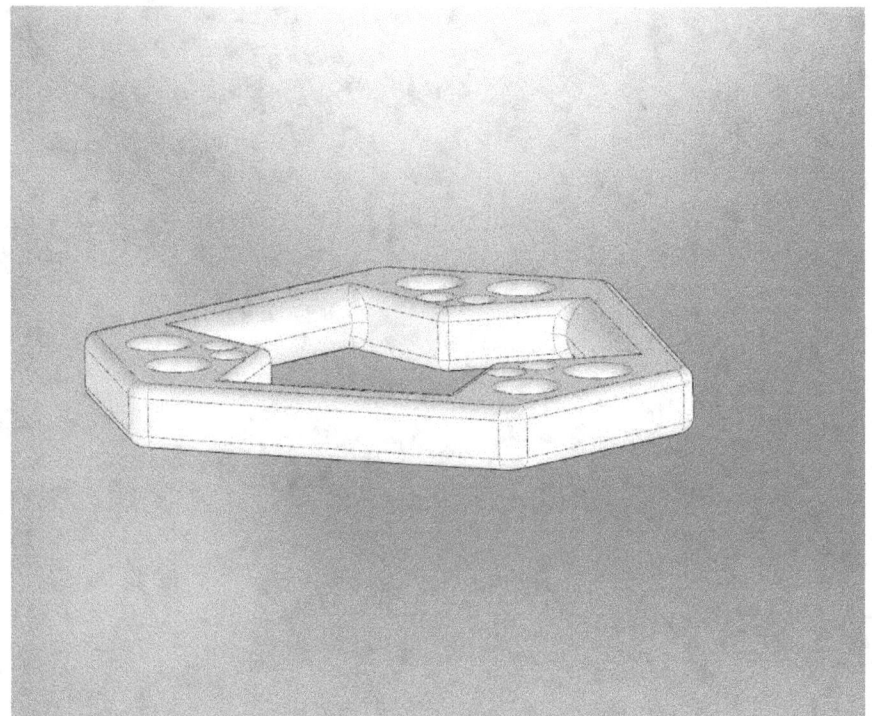

Figure 6.3-9. Rotation Torque Limiter Keeper for Main Harness Legs

For this architecture, the keeper was to be stowed in a bag in the forward bay, secured alongside the gusset between the two drogue mortars. The three main harness legs were rigged to pass through the keeper. When the three main parachutes were deployed, the keeper was held by straps at a position about 6 ft beneath the confluence fitting. This would provide a 6-ft section of gathered, parallel harness legs between the confluence fitting and the keeper. This section of gathered harnesses was envisioned to 'twist up' and 'untwist' in operation, in response to commands from RCS rockets (see Figures 6.3-10 and 6.3-11).

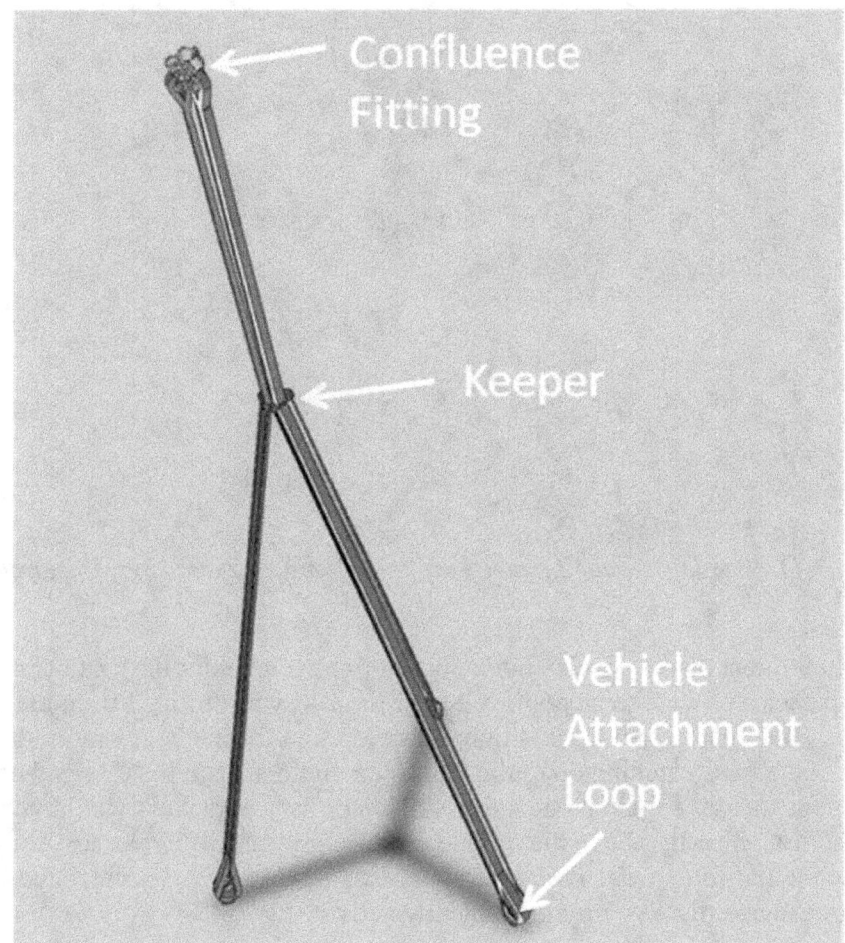

Figure 6.3-10. Rotation Torque Limiter Keeper for Main Harness Legs, Deployed

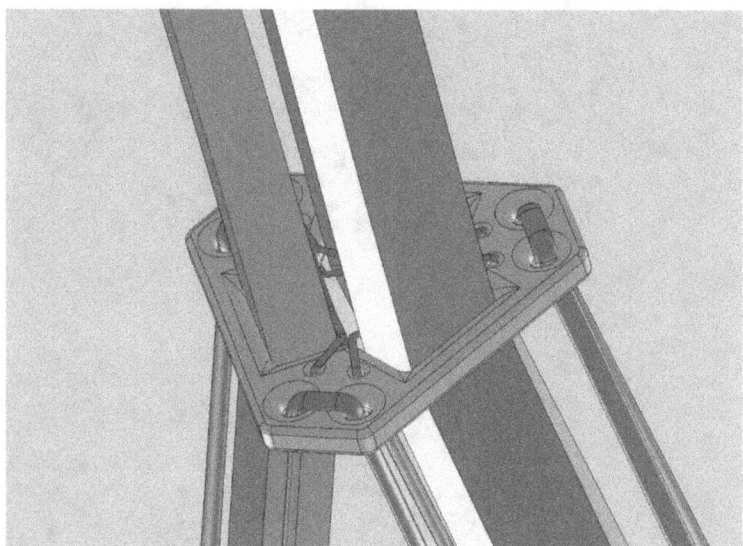

Figure 6.3-11. Rotation Torque Limiter Keeper for Main Harness Legs, Deployed Detail

During RP-1, the Project made a decision to add two lightweight auxiliary parachutes to the FBC design. These parachutes were to be deployed after the CM was dropped from the drogue-suspended FBC. This would provide additional deceleration for the heavy monolithic FBC, to reduce the risk of the FBC catching up to the main-parachute-suspended CM and contacting and damaging components of the main parachutes. Each auxiliary parachute would be stowed on the underside of the FBC, directly above the drogue mortars prior to FBC release. They would be deployed by the inertial force from a metallic slug attached by a cable, accelerated by a 'drogue gun' that could be mechanically or pyrotechnically activated.

The auxiliary parachutes were approximately 32-ft ringsail parachutes (see Figure 6.3-12).

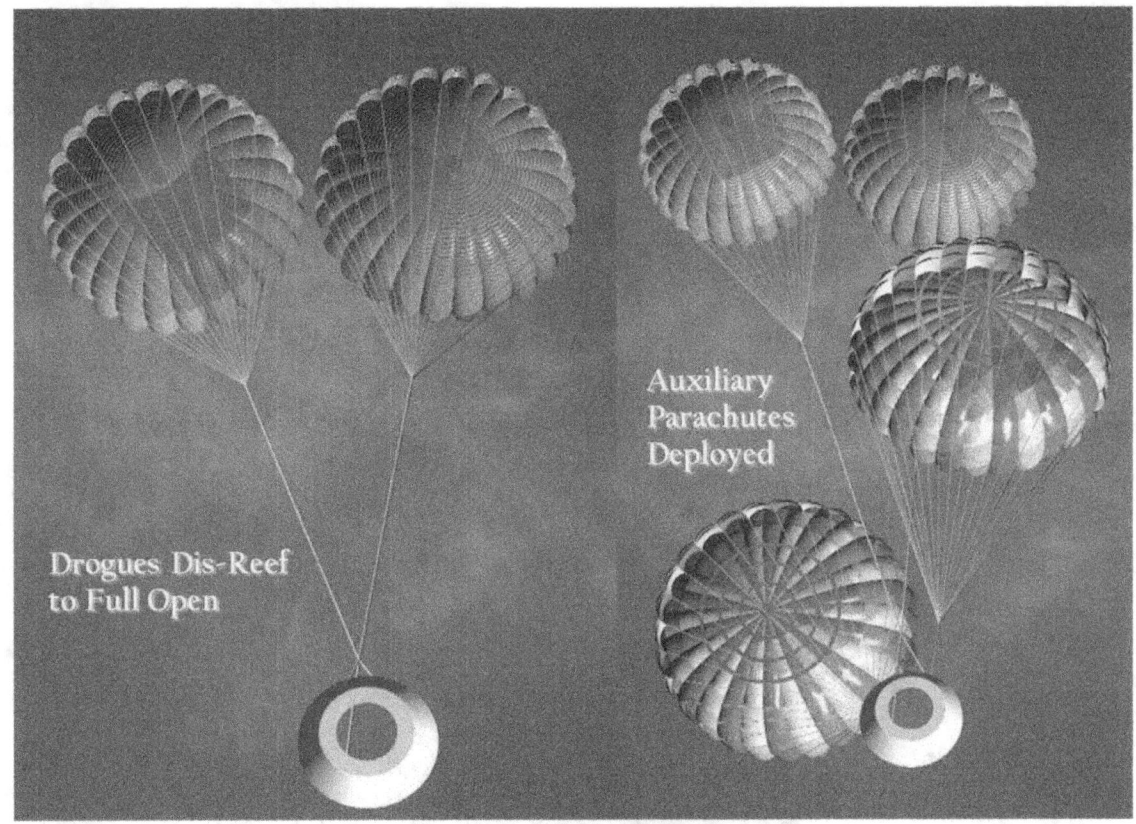

Figure 6.3-12. Auxiliary Parachutes

Near the end of RP-1, the Project made two more significant changes to the architecture.

In certain deployment circumstances, 'pre-stroking' of the textile energy modulators that were built into the d-bag handles was thought to pose a risk of burning or damaging of a main parachute during deployment. The addition of textile bootstrap devices called 'kicker straps' was considered as a possible solution but ultimately discarded by the Project.

The Project opted instead to delete the FBC extraction of the main parachute d-bags and add three FBC-extracted pilot parachutes to the architecture. The pilot parachute d-bags were to be secured with retention cords to the underside of the FBC, each one on the opposite side of a gusset from a main parachute pack (see Figure 6.3-13).

When the FBC was released and the CM separated from the drogue-supported FBC, the pilot d-bag restraints would each be cut by a knife on a textile lazy leg strap. The continued physical separation of the CM free falling away from the drogue-suspended FBC would then pull the pilot risers, suspension lines, and parachute canopies from within their respective d-bags. The pilot parachutes would open to full without dis-reefing. The pilot parachutes were approximately 10-ft-diameter conical ribbon cut-gore parachutes.

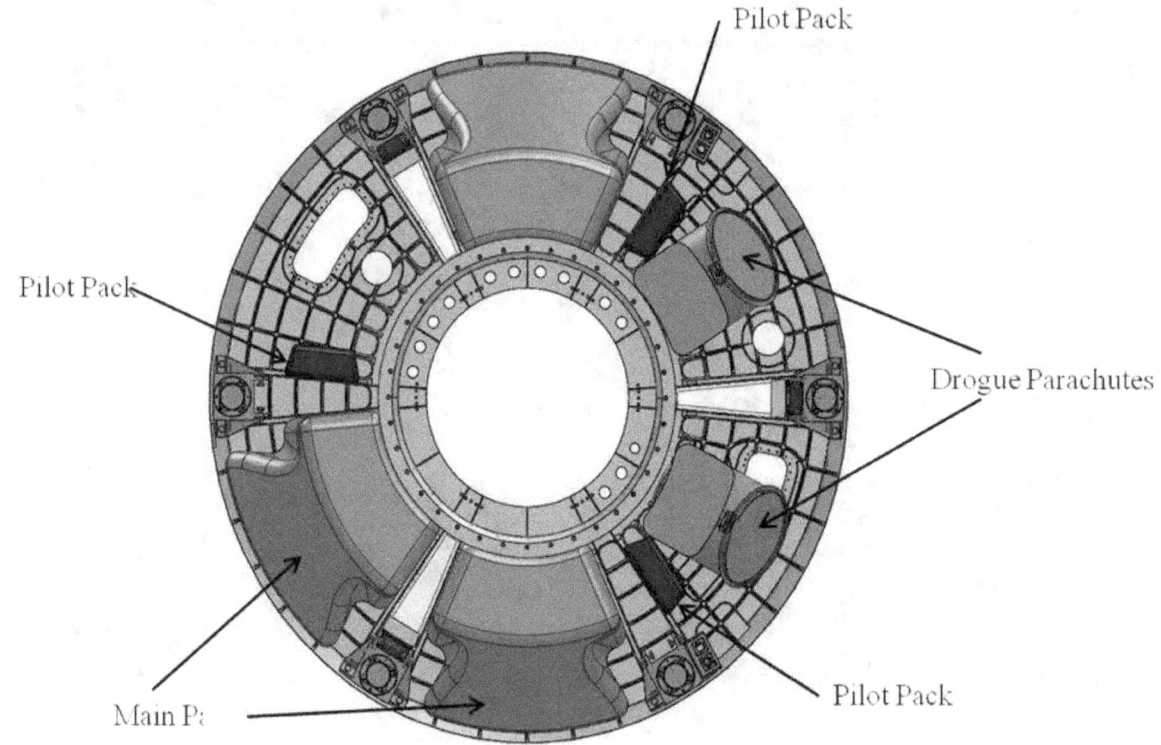

Figure 6.3-13. Pilot Parachutes Stowed

Once opened, each pilot parachute riser would then propagate the unlacing of the retention system on one main d-bag and lift it from its stowed location (see Figure 6.3-14).

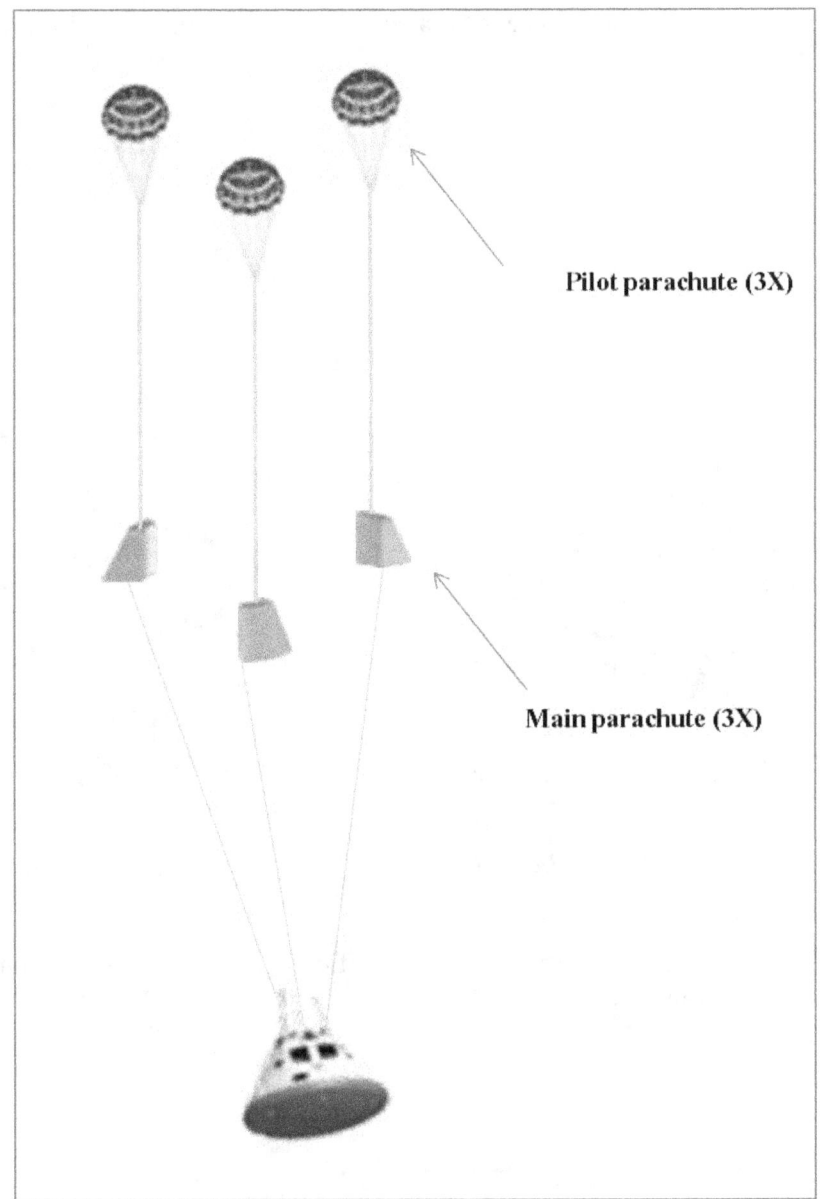

Figure 6.3-14. Pilot Parachutes Deployed

The Project also replaced the single drogue confluence fitting with two individual drogue confluence rings. The metallic rings would be stowed, one each, inside the drogue mortars on top of the drogue parachutes and risers. With this change, three textile harness legs were replaced with six thinner textile harness legs, stowed as before on top of the FBC and covered with TPS (see Figure 6.3-15).

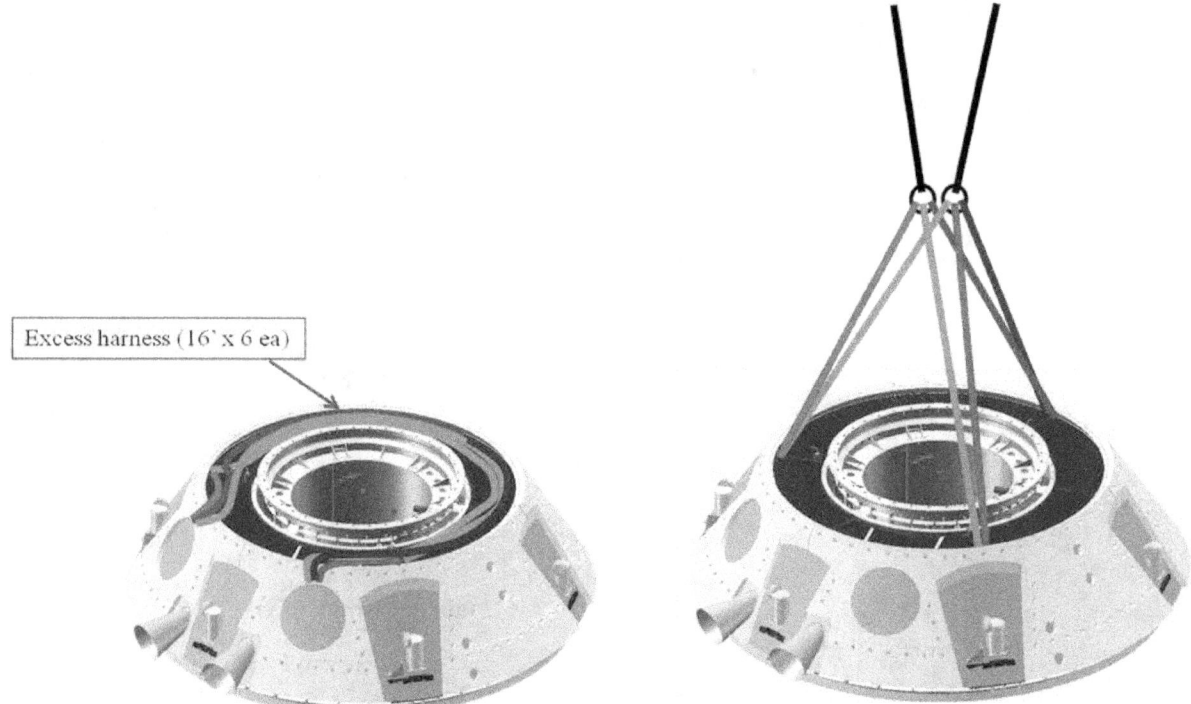

Figure 6.3-15. Drogue Harness Legs Stowed and Deployed, with Dual Confluence Rings

The three drogue harness leg attach fittings were modified to accommodate two harness ends each, instead of one (see Figure 6.3-16).

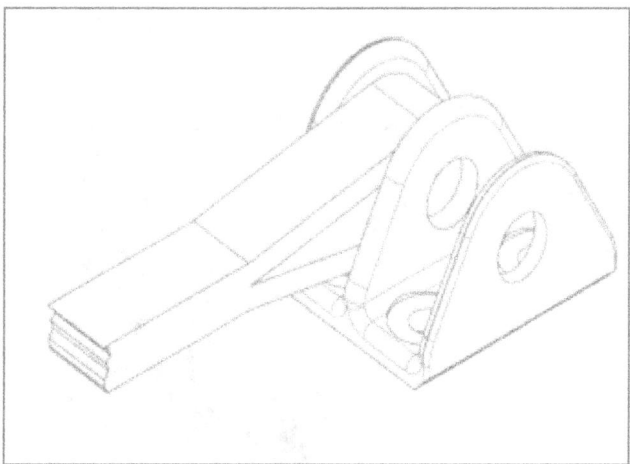

Figure 6.3-16. Drogue Harness Leg Attachment Fittings Detail

6.3.2 Review Period 2—April 2009 through August 2009

RP-2 spanned April 2009 through August 2009. Early in this period the CPAS PDR was rescheduled from June to September 2009.

Early during RP-2, the Project changed the attachment scheme for the main parachutes. The main parachute confluence fitting and the three main parachute harness legs were eliminated. The three main risers were lengthened accordingly and their ends were all three to be attached to the "zero-" gusset (i.e., the CM forward bay gusset that was between the two drogue mortars). The riser fairlead (i.e., the riser guide integrated into the gusset) was first envisioned as a triangular hole of similar geometry as the keeper in the top sloped surface of the gusset (see Figures 6.3-17 and 6.3-18).

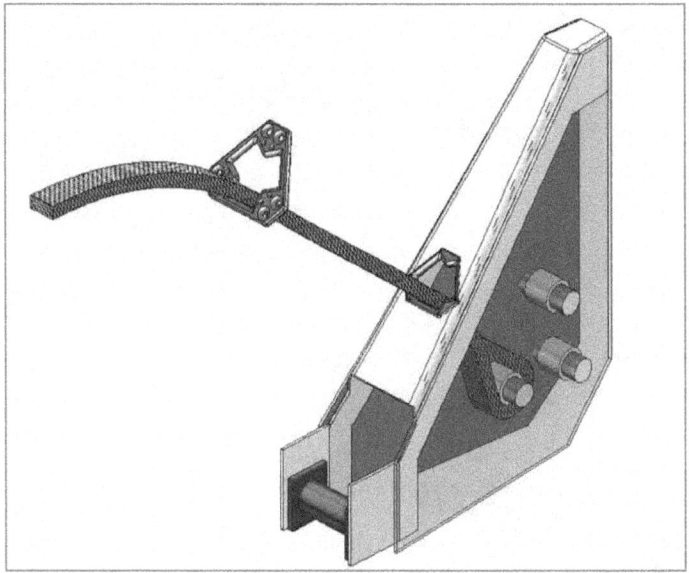

Figure 6.3-17. Zero-gusset for Main Risers Single Attach Point, First Iteration

Later in RP-2, the fairlead was revised to be three aligned, circular guide holes in the top sloped surface of the gusset.

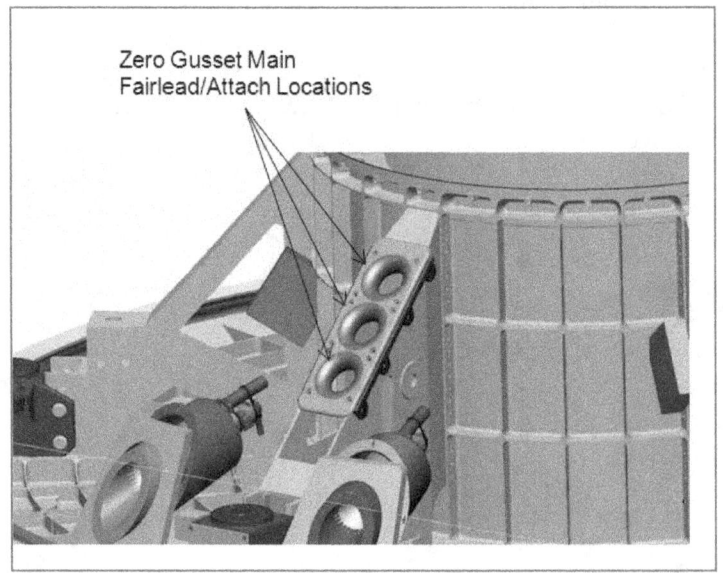

Figure 6.3-18. Zero-gusset for Main Risers Single Attach Point, Second Iteration

For this architecture, the metallic keeper was retained and was still to be stowed in a stowage bag in the forward bay, alongside the gusset between the two drogue mortars (see Figure 6.3-19).

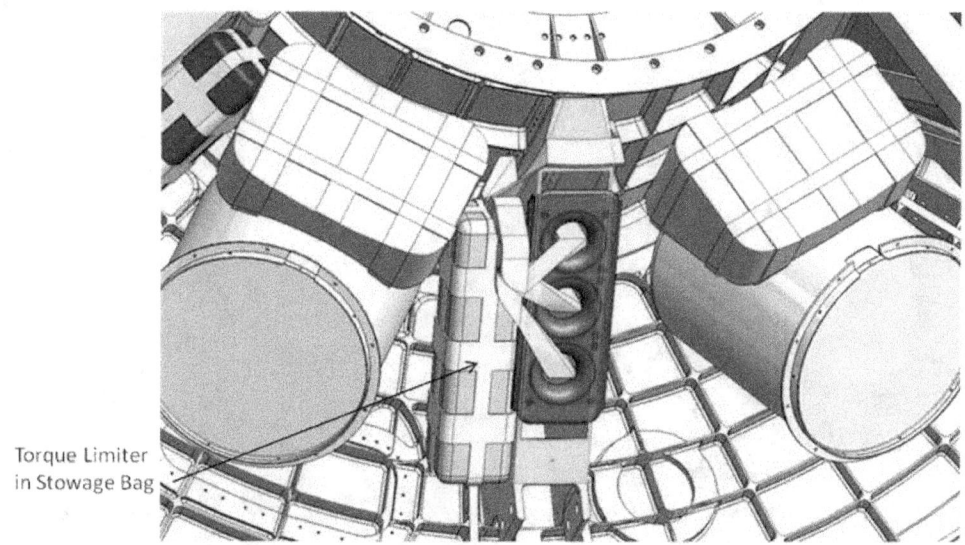

Figure 6.3-19. Rotation Torque Limiter Keeper for Main Risers, Stowed

In this application, the three main risers would be rigged to pass through the keeper, rather than the harness legs (as described in Section 6.3.1). The keeper would be held at the appropriate deployed position along the gathered risers by beckets sewn into the risers (see Figure 6.3-20). In this architecture, the length of the three gathered risers between the zero-gusset and the keeper was envisioned to 'twist up' and 'untwist' in operation, in response to commands from RCS rockets. This would allow the roll maneuverability to orient the CM just prior to water landing.

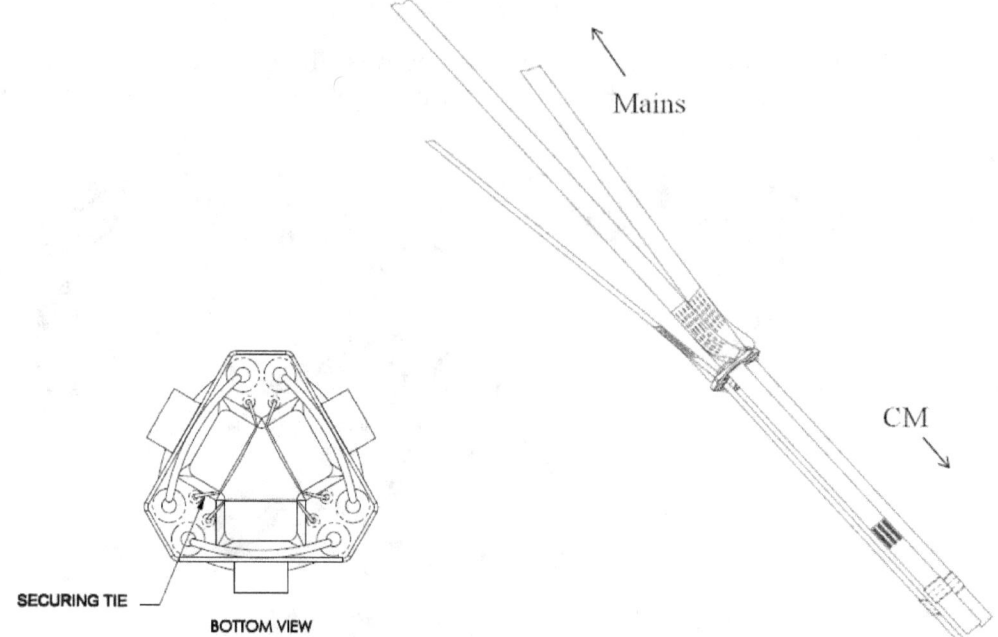

Figure 6.3-20. Rotation Torque Limiter Keeper for Main Risers, Deployed Detail

The Project conducted IDR-2 in May 2009, at which several additional architectural modifications were identified, including revision of the main parachute d-bag shape and the main d-bag retention and release system. Each main d-bag would be held in place by an upper and a lower textile retention panel. Each panel interfaced with the d-bag along one outer edge, using a daisy-chain lacing feature that had heritage design in the Apollo Program ELS main parachute daisy-chain retention system (see Figures 6.3-21 and 6.3-22).

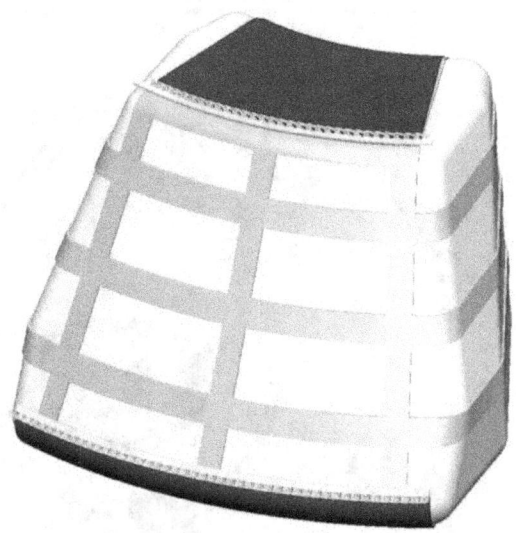

Figure 6.3-21. Main Pack Retention System at IDR-2

Figure 6.3-22. Main Parachute Pack Retention System, Detail

At IDR-2 it was learned that the Orion Project changed the configuration of the CM Uprighting System (CMUS) tanks, replacing a single tank with five helium-filled composite overwrapped pressure vessels (COPVs). The CMUS system would be utilized in the event the CM became inverted following a water landing. It was noteworthy here because the system's five uprighting bags, five COPVs, and associated plumbing occupied the CM forward bay and so were integrated with CPAS components. The five CMUS inflatable uprighting bags were stowed behind the three main parachutes, against the CM tunnel wall (see Figure 6.3-23).

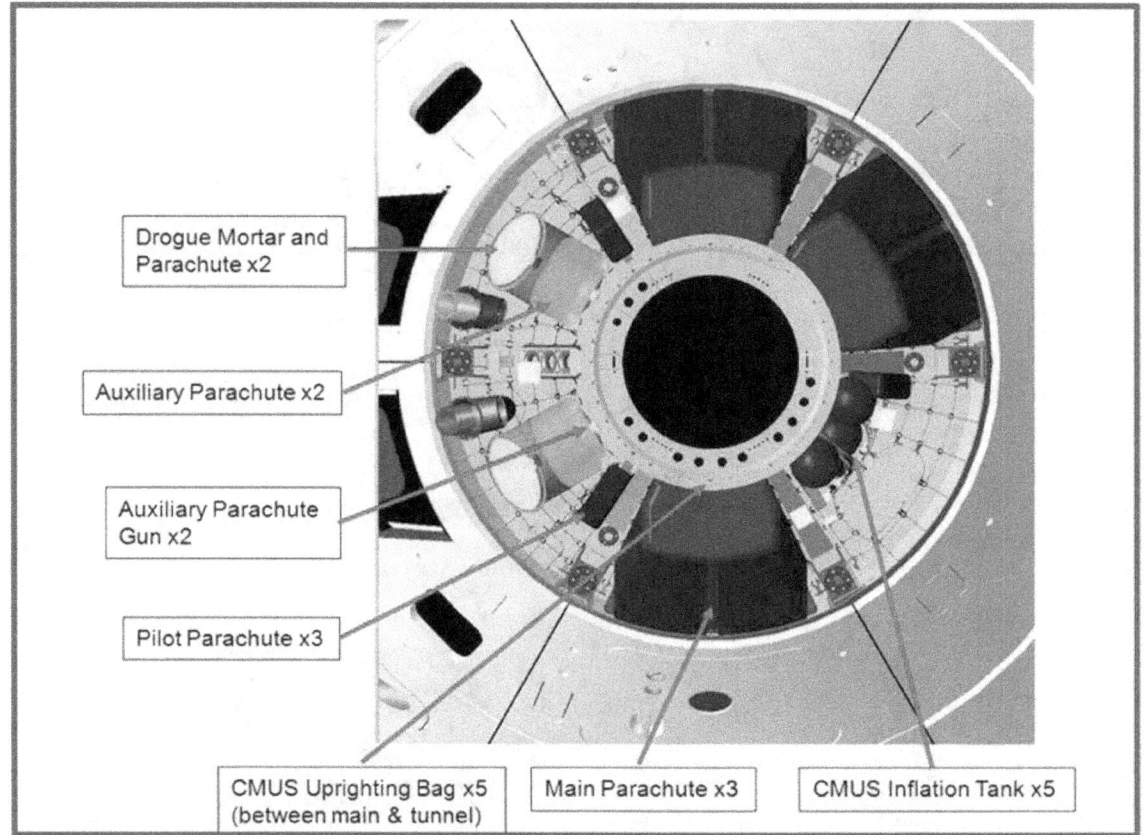

Figure 6.3-23. CM Forward Bay Arrangement at IDR-2, Detail

In June 2009, the Orion Project chartered an Integrated Design Assessment Team (IDAT) that included CPAS personnel, contractor and NASA personnel with responsibility for systems with which CPAS interfaced, NASA and contractor discipline experts, and NASA personnel from the Mars Science Laboratory EDL design development team. The IDAT was chartered to assess sufficiently mature integrated design options for CPAS and to make design recommendations to the ERB for adoption by the Orion Project (see Figure 6.3-24).

The IDAT was tasked to assess six key CPAS design issues:

1. Insufficient parachute packaging volume and high packing densities
2. Risk of parachute damage from near field re-contact during FBC separation
3. Large mass threats due to high drogue inflation and kick loads
4. Risk of failure of drogue deployment due to drogue harness extraction through TPS

5. Risk of parachute damage from far field re-contact of FBC due to immature concept for auxiliary parachute deployment
6. CPAS as a top risk contributor to LOC for the Orion Project

The IDAT explored four integrated areas of design trade space for a more reliable CPAS:

1. FBC design options and methods of its jettison
2. LAS design options and methods of its jettison
3. CPAS parachutes methods of extraction and deployment
4. FBC design options that may increase available volume for CPAS components

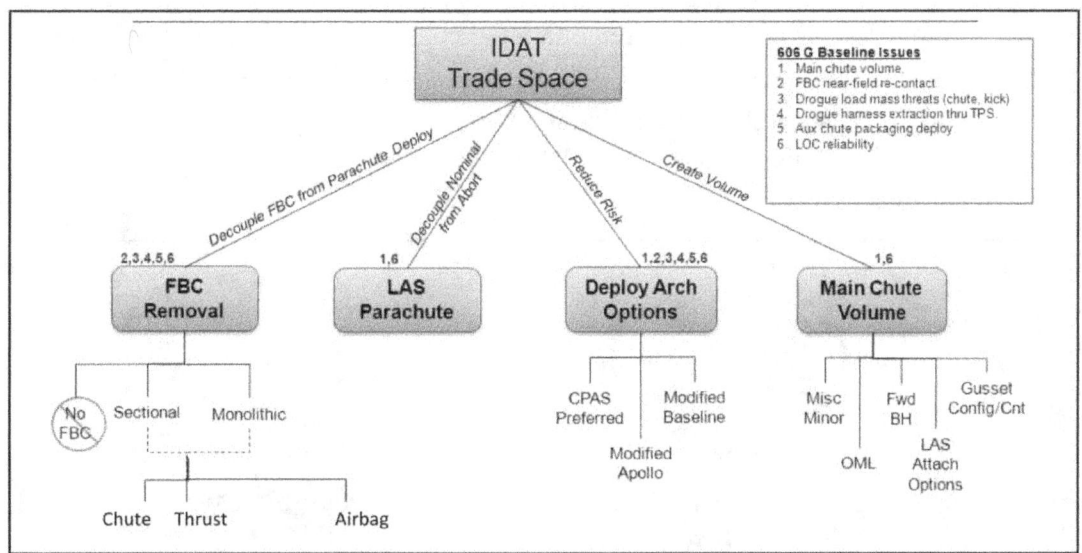

Figure 6.3-24. Overview of IDAT Charter

6.3.3 Review Period 3—September 2009 through April 2010

RP-3 spanned September 2009 through April 2010. Near the end of RP-2, the CPAS PDR had been rescheduled from September 2009 to March 2010.

The Project conducted IDR-3 in September 2009. At this review, the IDAT activity was ongoing so the review focused largely on Project Technical Requirements Specification (PTRS) Rev. A, April 2009, JSC-63497 [Ref. 7]; the Companion Document Rev. B, May 2009, JSC-64355 [Ref. 8]; and the Master Verification Plan (MVP) approach framework.

In November 2009 the Orion Project and CPAS adopted a number of architectural recommendations from the IDAT for CPAS, CM, and some of its other subsystems. They are described in the following sections, along with additional design refinements incorporated before April 2010.

The CPAS Con Ops was changed such that FBC jettison would become the initiating event for atmospheric reentry, or following a pad or launch abort. In addition, the FBC design was changed from a single large structure to a segmented design comprised of six individual L-shaped panels, sealed where they meet over each gusset. Each panel was to be held in place by two retention-and-release (R&R) mechanisms, one to fasten its top surface to the CM tunnel and one to fasten its bottom edge to the CM bulkhead (see Figures 6.3-25 and 6.3-26).

All six panels would be released from the CM at subsonic speed, by redundant activation (by electrical or ordinance) of one length of mild detonating cord (MDC) that travels 360 degrees around the CM along the panels' lower circumferential edges, and one length of MDC that travels 360 degrees around the CM tunnel along the panels' upper circumferential edges.

Each panel would be jettisoned by rapid inflation via redundant gas generators, of a single circular-section airbag installed beneath its larger panel surface.

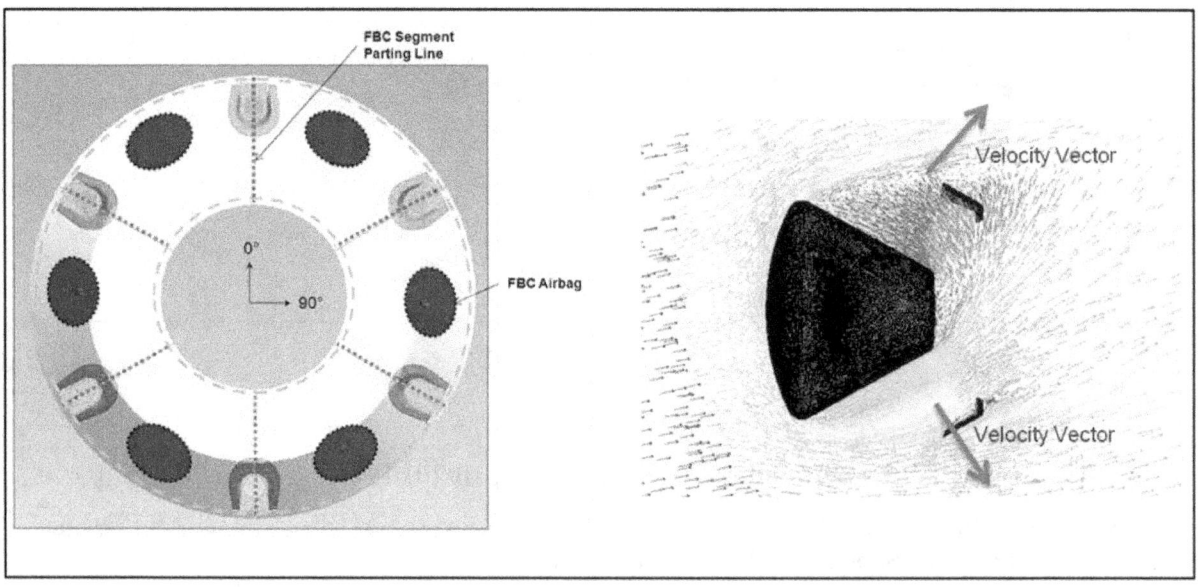

Figure 6.3-25. Segmented FBC and Representative Flow Analysis Image

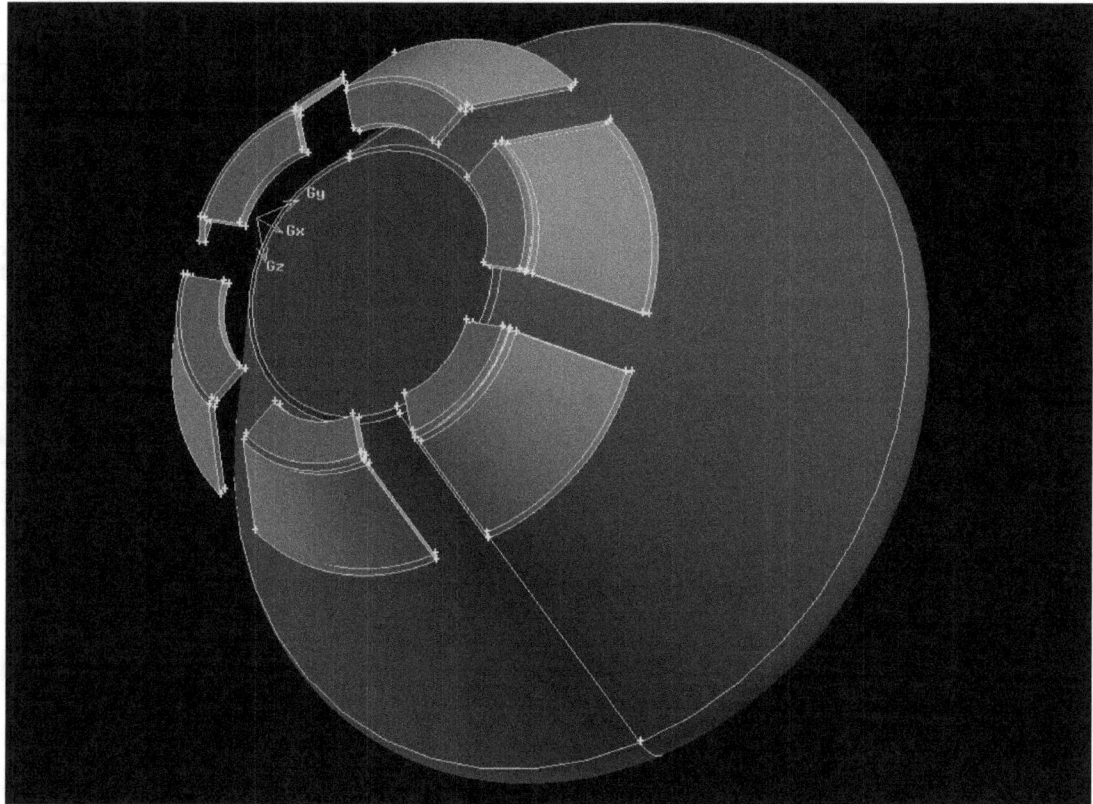

Figure 6.3-26. Segmented FBC Detail

In three forward bay sectors, the airbag force would react against a main parachute pack. In two others, it would react against a drogue mortar cover. In the sixth sector, the airbag would react against a newly added reactive structure installed over the CMUS COPVs (see Figure 6.3-27).

The auxiliary parachute system that had been previously included to lower the ballistic coefficient of the monolithic FBC after its release was deleted. Each FBC panel would retain its inflated airbag when jettisoned to provide it with a lower ballistic coefficient during descent.

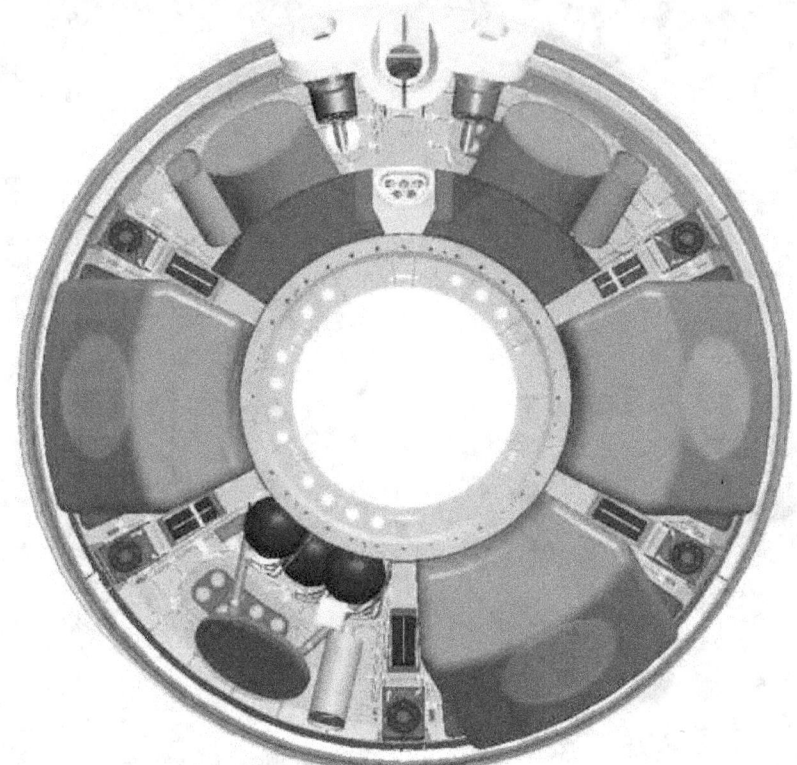

Figure 6.3-27. CM Forward Bay Detail Showing FBC Panel Circular Push-off Areas

The FBC-attached harnessed-drogue parachute architecture was replaced with an architecture where both drogue risers were attached inside the CM's zero-gusset, along with the three main risers. The fairlead with five riser guide holes was referred to as the 'flowerpot.' The two drogue parachutes and risers were stowed and deployed from within mortars, as with the previous architecture, but the two metallic drogue confluence rings were deleted. The drogue risers were changed from textile to steel cables. Lengths of steel riser were to be coiled and stowed inside the mortars. Significantly, the drogue risers were routed under the FBC and did not interact with the FBC panels. (See Figure 6.3-28.)

Structural elements were added to span the zero-gussets and the two gussets on either side, so that all three gussets would participate in reacting riser loads across the predicted range of riser angles. Drogue riser cutters were contained within the thick zero-gusset.

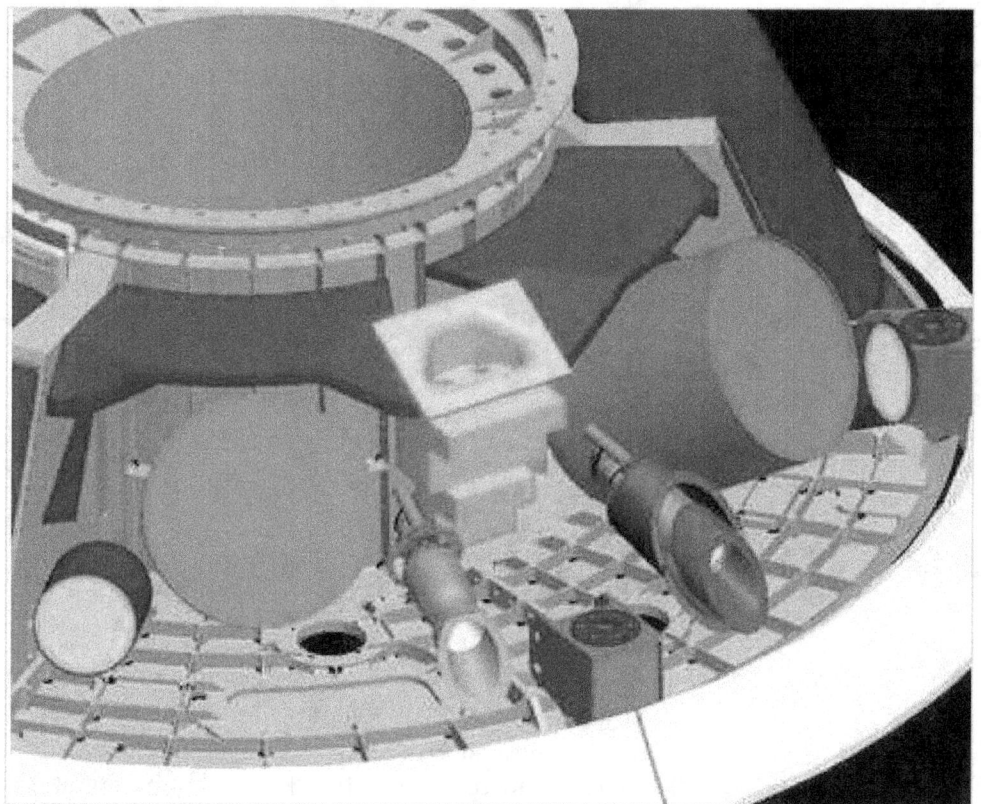

Figure 6.3-28. CM Zero-gusset Riser Flowerpot Fitting

The five gussets other than the zero-gusset were redesigned to package the five CMUS uprighting bags, thereby removing the bags from being stowed behind the stowed main parachutes. The gussets' redesign also accommodated stowage of parachute risers across their top surfaces, and routing of CMUS plumbing beneath them. They also incorporated features along their edges for integrating with the main parachute d-bag retention system. (See Figure 6.3-29.)

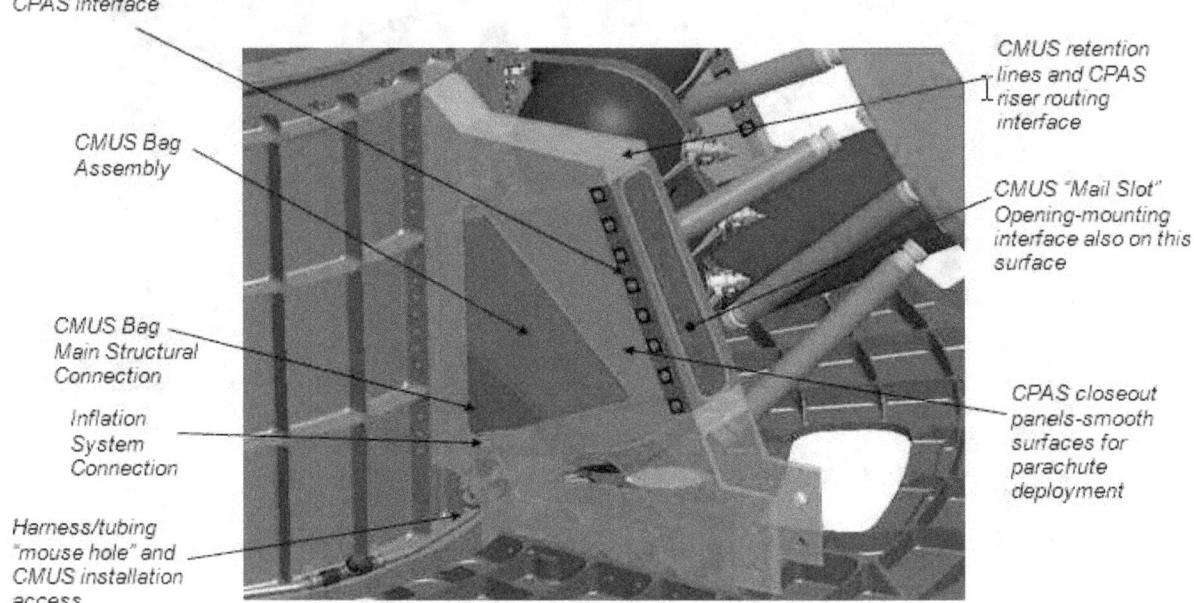

Figure 6.3-29. CM Common Gusset

The CPAS Con Ops was changed such that very shortly following FBC panels jettison, the drogue parachutes were deployed only for atmospheric reentry. For pad or ascent aborts up to a to-be-determined altitude, the drogues would not be used. Rather, pilot parachute deployment would follow quickly after FBC panels' jettison, to shorten the time between FBC jettison and landing. In either case, panel jettison was a required precursor to mortar firing. No means was described to liberate a FBC panel stuck over a mortar.

To provide more volume in the CM forward bay and to maintain main parachute packed densities below the required 38 psf, the Orion Project increased the vehicle backshell angle by 2.5 degrees. This was accomplished principally with changes to the backshell TPS; the pressure vessel structure below the forward bay was minimally affected (see Figure 6.3-30).

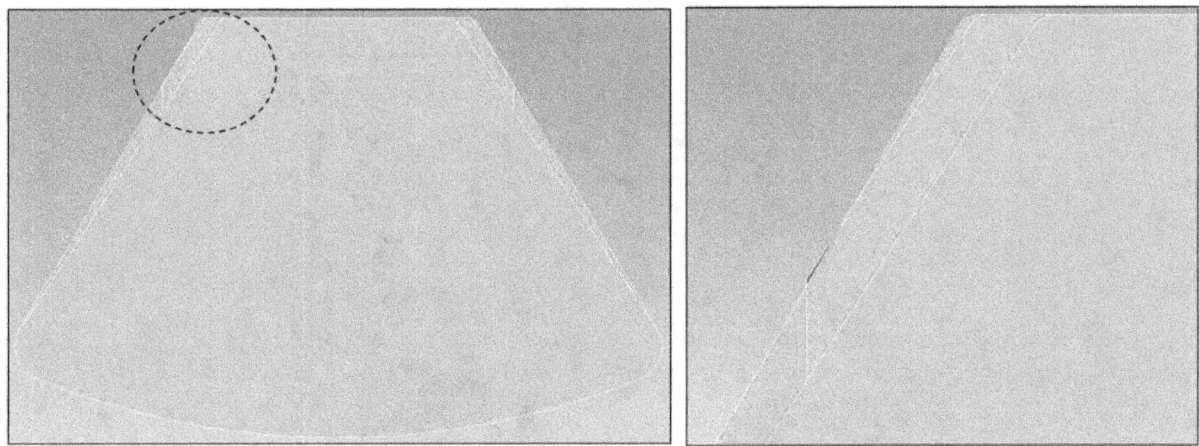

Figure 6.3-30. CM Backshell Angle Increase

In addition, the integrally machined ribs on the CM bulkhead that serves as the floor of the forward bay were reduced in height in the three sectors where main parachute d-bags are stowed. Changes were also made to the LAS well design. With balanced consideration given to various risks of LAS contact with the FBC during LAS jettison, the LAS wells were moved 2.3 in. radially outward and the LAS well escape angle was changed to 10 degrees with a 1-in. dynamic clearance. Some LAS/CM electrical connectors were also relocated from the forward bay into the LAS.

The main parachute deployment architecture was changed significantly during RP-3. Mortar-deployed pilot parachutes replaced the FBC-deployed pilot parachutes. Three pilot mortars were located on opposite sides of gussets immediately adjacent to the three main parachutes. This resulted in one pilot mortar in each of the two sectors that already held a drogue mortar and one pilot mortar in the sector that contained the CMUS COPVs. Two variants of the pilot mortars can be seen in the detail in Figures 6.3-27 (racetrack-cross-section) and 6.3-28 (circular-cross-section).

The main parachute d-bag retention system was redesigned twice during RP-3. Early in this review period, the Project redesigned the main d-bag retention system to include a daisy-chain arrangement of locking loops holding four textile retention flaps closed over the main d-bags. This retention scheme can be seen in Figure 6.3-31. Note in the detail that this arrangement preceded the change from FBC-deployed pilot parachutes to mortar-deployed pilot parachutes.

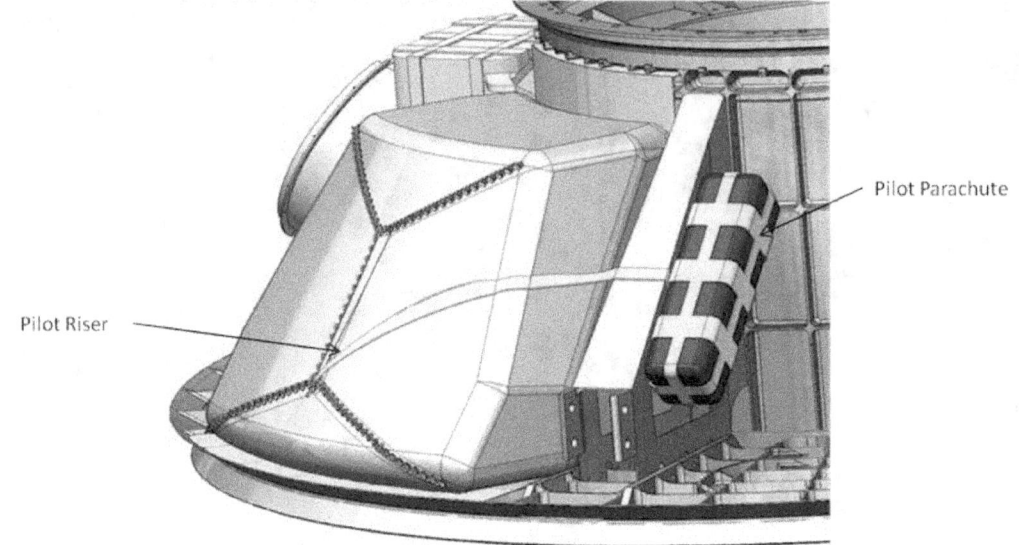

Figure 6.3-31. Main Pack Retention System

Unzipping of the d-bag retention loops was to have been initiated by breaking a break tie on each pack, by the action of a cut-knife pulled by a lanyard attached to the pilot riser (see Figures 6.3-32 and 6.3-33). The continued action of the main riser would subsequently lift the d-bag from its stowed location beneath the four flaps.

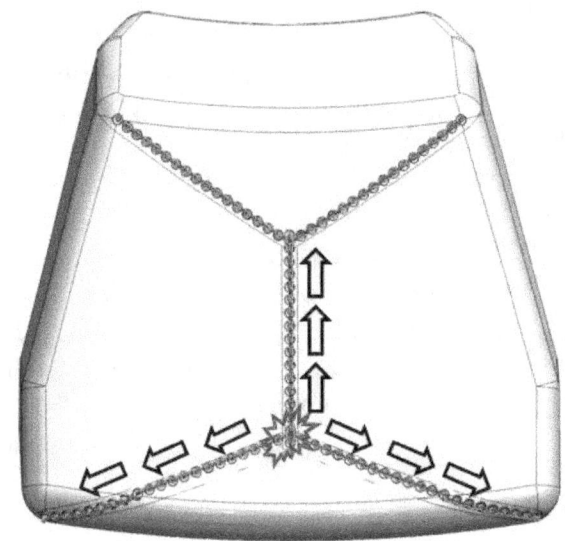

Figure 6.3-32. Main Pack Retention System, Detail 1

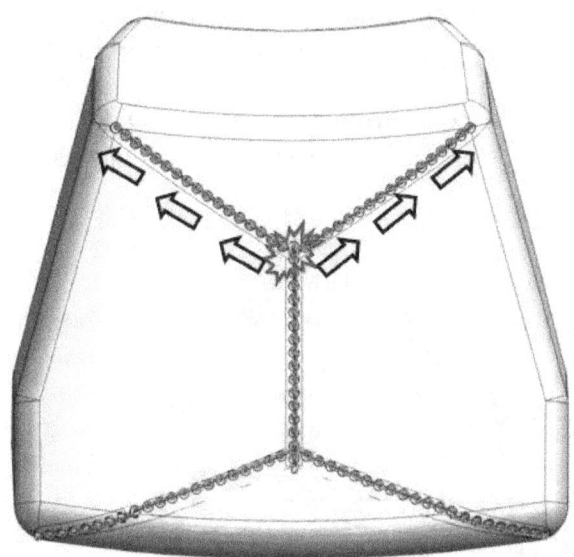

Figure 6.3-33. Main Pack Retention System, Detail 2

A system of corsets and beckets would be cinched during rigging to help the d-bag retain its shape and prevent its contact with the FBC inner mold line (IML) (see Figure 6.3-34).

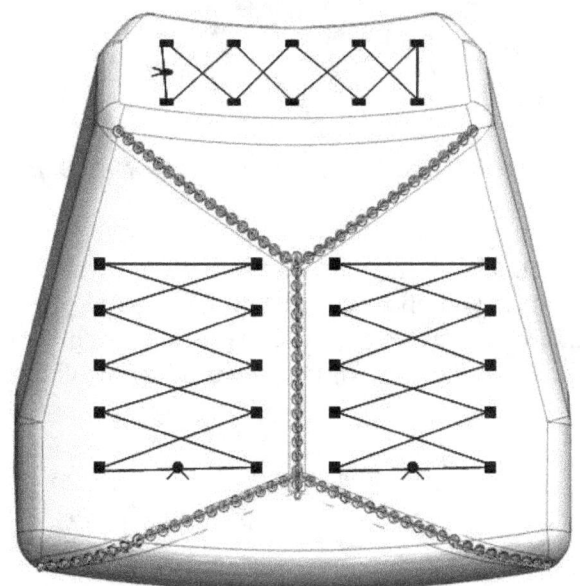

Figure 6.3-34. Main Pack Retention System, Detail 3

Later during RP-3, in part to utilize the additional volume provided by the increased CM backshell angle, the main d-bag's outer surface shape was simplified to a large-radiused surface with no facets. Its upper outer and lower outer edges were designed to be a constant radius from gusset to gusset. Its shape was to be controlled by a system of corsets and beckets sewn into the d-bag (see Figures 6.3-35). As with the previous d-bag design, these would prevent contact with the FBC IML.

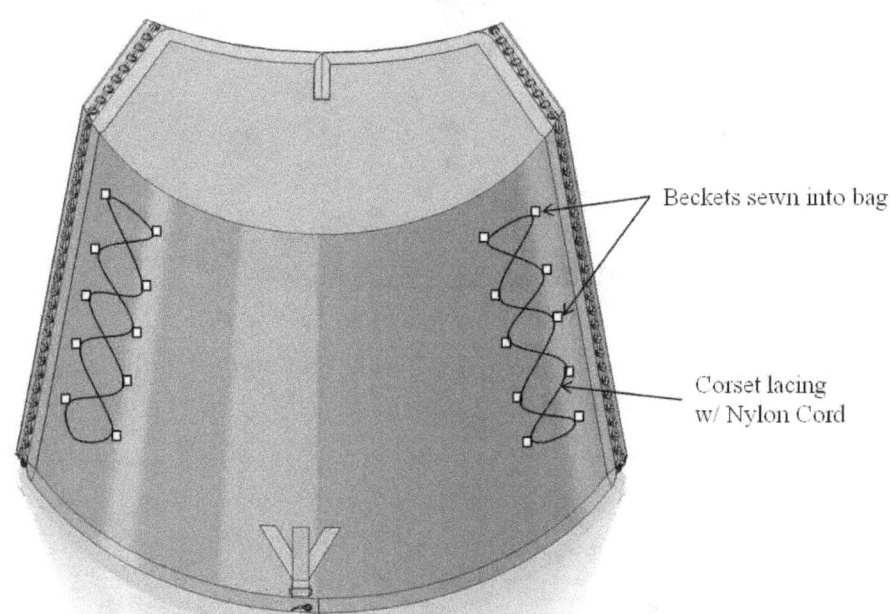

Figure 6.3-35. Main Pack Retention System, Detail 4

Each main d-bag was to be retained to the two gussets between which it was stowed using mated textile panels on the gusset outer and upper edges, and the d-bags. The panels were looped together using a daisy-chain retention system. A crow's foot along the outer lower edge of the d-bag pinned to an attach strap on the floor of the forward bay provided additional restraint (see Figure 6.3-36).

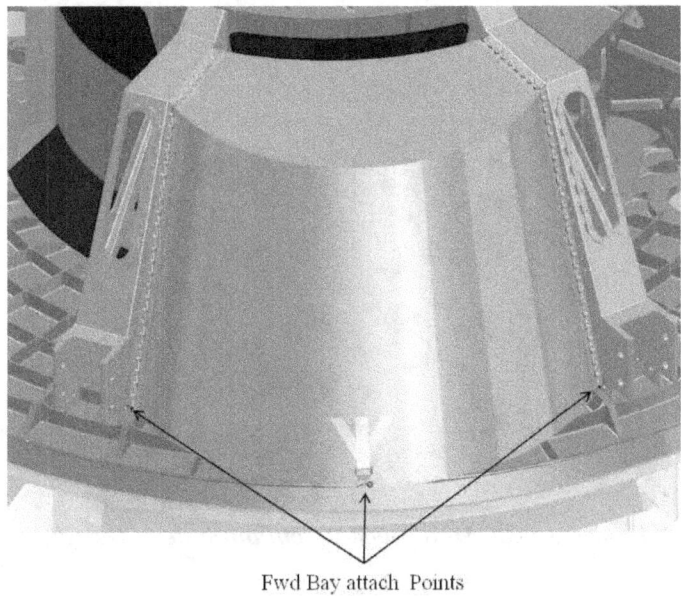

Figure 6.3-36. Main Pack Retention System, Detail 5

One curve pin terminated each of the two daisy chains and a third pin secured the crow's foot. Lanyards were attached to each textile pilot parachute riser to pull these curve pins out and initiate the unlacing of the daisy chain loops. (See Figures 6.3-37 and 6.3-38.)

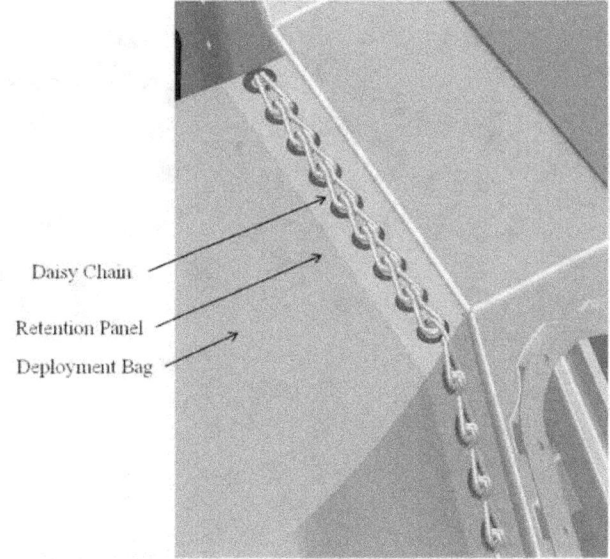

Figure 6.3-37. Main Pack Retention System, Detail 6

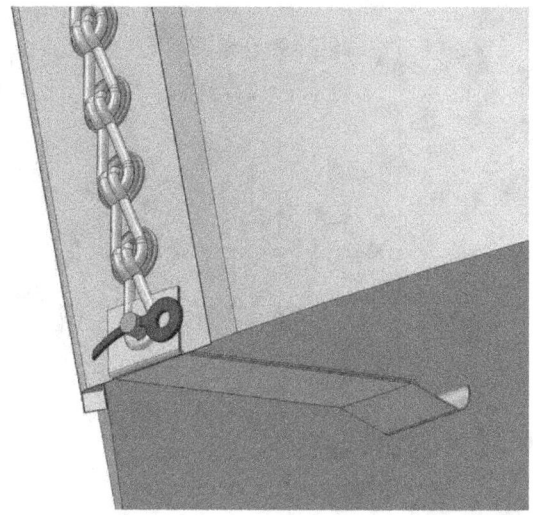

Figure 6.3-38. Main Pack Retention System, Detail 7

The pilot risers would split along this span, and the two textile bridles would each unzip one side of retention loops (see Figure 6.3-39). The assessment team's understanding was that the riser halves would be integrated loop-by-loop along the length of the daisy chain, to assist its opening. The two textiles would rejoin as one textile riser on the pack top. Continuing action by the pilot parachute force would lift the released main d-bag from its stowed position at a central lift point in the middle of the d-bag top surface.

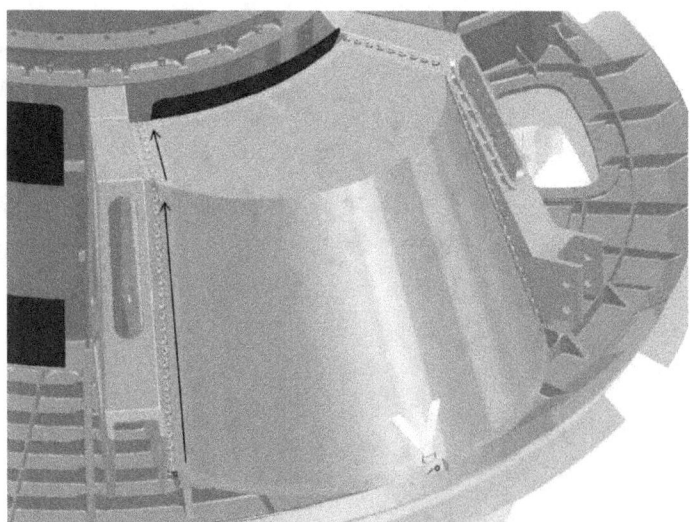

Figure 6.3-39. Main Pack Retention System, Detail 8

The main parachutes, suspension lines, and most of the risers were still to be packed into the main d-bags. The main risers were changed from textile to braided steel cables. Various design options for cutting the steel drogue and main risers inside the zero-gusset were being vetted at the end of RP-3.

The metallic torque reducing keeper remained as a part of the architecture at the end of RP-3. The Project was conducting analysis and planning tests to determine whether the closeness of the riser guide holes in the flower pot was sufficient to limit the roll torque to the required maximum. If it became necessary to retain the keeper in the design, then it and the lengths of steel main risers between the zero-gusset and the keeper would be stowed in the FBC.

At the end of RP-3, the CM also included two ramps in the forward bays behind each main parachute. The lower ramp provided enclosed space for routing electrical and CMUS plumbing around the base of the CM tunnel. The upper ramp served to protect the main d-bag from effects of the detonation of the upper MDC that was to release the FBC panel. (See Figure 6.3-40.)

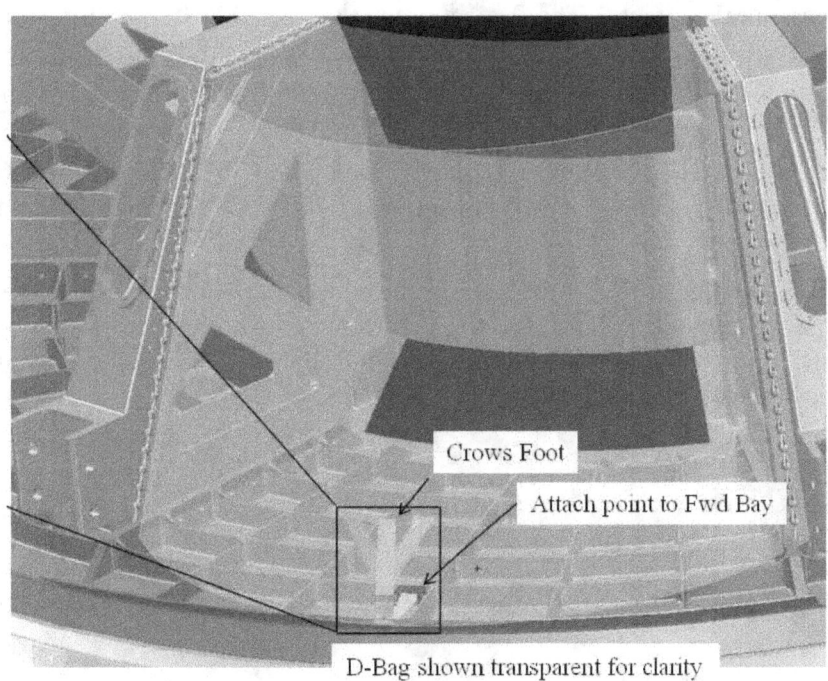

Figure 6.3-40. Main Pack Ramps

6.4 PRA and other Reliability Products

The evolution and refinement of the Project's PRA and other reliability information are described in the following sections.

6.4.1 Review Period 1—December 2008 through March 2009

A preliminary FTA, Hazard Analysis (HA), and PRA were provided at IDR-1. PRA failure estimates were generated from reliability data obtained from the following sources:

1. Parachute drop data from the U.S. Army, Apollo Program, Soyuz Program, and SSP SRB.
2. Vendor component data.
3. Component reliability predictions computed from the intersections of material strength and applied load probability distributions (load-strength interference method).

The Project showed data that SSP SRB parachute contact with its main parachute support system during extraction had occurred 3 times out of 651 demands (1:217). The CPAS Project assessed that this data was pertinent to CPAS, equivalent to the CPAS failure mode "deployment failure damage due to structural contact." The Project noted that the SSP SRB architecture assured that only one out of three parachutes could fail by this mode, but that the CPAS architecture at that time did not preclude multiple parachute contacts with structure. The Project computed system reliability if only one CPAS main parachute failed by this mechanism to be 1:8,205.

The failure estimate of 1:217 became the driving value in the CPAS PRA. The Project noted that simulation analyses and ground and flight testing would ultimately refine this failure probability for the CPAS architecture.

AS provided reliability data at IDR-1 for the drogue and main parachute assemblies. The reliability of most components had been *predicted* from the intersections of material strength and applied load distributions to determine the "areas of unreliability" (from the area of intersection of the two distributions). (See Tables 6.4-1 and 6.4-2.)

Table 6.4-1. Drogue Parachute Reliability Estimates at IDR-1

Description	Reliability Prediction	Source	Failure Rate (1/x)
Drogue Parachute	0.9966997	Sum	2,473
Parachute Packing	0.9999000	Test	10,000
Drogue Bag Separation	0.9997000	Test	3,333
Suspension Lines/Skirt Attach	0.9999999	Prediction	10,000,000
Vent Line Attach	0.9999999	Prediction	10,000,000
Suspension Lines/Riser Attach	0.9999999	Prediction	10,000,000
Drogue Riser	0.9999993	Prediction	1,428,571
Horizontal/Radial	0.9999999	Prediction	10,000,000
Skirt Band	0.9999999	Prediction	10,000,000
Reefing Line	0.9999999	Prediction	10,000,000
Cutter Operation, Drogue Flap	0.9999999	Prediction	10,000,000

Table 6.4-2. Main Parachute Reliability Estimates at IDR-1

Description	Reliability Prediction	Source	Failure Rate (1/x)
Main Parachute	0.9956148	Sum	568
Parachute Packing	0.9970000	Test	10,000
Quarter Bag Attach	0.9999999	Prediction	10,000,000
Suspension Lines/Skirt Attach	0.9987597	Prediction	806
Vent Line Attach	0.9998816	Prediction	8,446
Suspension Lines/Riser Attach	0.9999998	Prediction	5,000,000
Main Riser	0.9999998	Prediction	5,000,000
Main Seam	0.9999998	Prediction	5,000,000
Skirt Band	0.9999999	Prediction	10,000,000
Reefing Cutter	0.9999700	Test	33,333
Reefing Line Strength	0.9999999	Prediction	10,000,000
Pin / Cutter Operation, Main Flap	0.9999999	Prediction	10,000,000

The Project summarized that individual components, such as a skirt band, typically are reliable based on their stress-strength calculations. The CPAS Project concluded that the intrinsic reliability of CPAS was limited by the main parachute reliability, which was limited by the reliability of the suspension lines and skirt attachment. This failure mechanism was noted by the Project to be amenable to design modifications. AS added that having one redundant drogue parachute and one redundant main parachute provided failure tolerance against catastrophic failure, and that the drogue and main parachutes were being designed to survive loading conditions associated with a skipped reefing stage.

In assessment team discussions during RP-1, reliability subteam members led productive discussions of reliability budget methods and benefits.

In February 2009, the assessment team conducted a meeting with the Project. The assessment team provided an advance list of 42 questions to improve its understanding of CPAS, its function, and associated reliability risks. The assessment team asked questions on the verification and validation (V&V) of the simulation codes used for design and reliability decisions, which was declared to be outside the scope of CPAS's responsibilities. The Project satisfactorily answered the assessment team's questions.

The same month, the Project briefed the assessment team on two separate versions of CPAS fault trees (FTs) it was maintaining. A comprehensive FT was being maintained by a NASA civil servant and simpler FT was being maintained by a Jacobs ESCG counterpart. The latter FT was to be linked to the evolving PRA. At this meeting the Project noted dashed lines in the FT that connected to faults key to CPAS reliability but considered administratively outside of CPAS. These were the responsibilities of developers of other CM subsystems. At this meeting the Project also identified where additional data still was needed to refine the PRA.

6.4.2   Review Period 2—April 2009 through August 2009

At IDR-2 the Project presented a status summary of its Risk Assessment Executive Summary Report (RAESR), failure modes and effects analysis (FMEA)/critical items list, FTA, and HA. These deliverable products were to be completed by the S&MA team prior to the Project's PDR. The PRA was described as a living work product subject to change from testing and analyses results. Two methods were described in use for determining PRA reliability values: predictions and estimates. Predictions were described as being based on historical or vendor data. Estimates were described as being based on expert opinion and engineering judgment (EJ). The Project indicated intent to conduct an EE exercise.

The Project identified 17 top-level failure modes as CPAS PRA 'input values' for the FTA (see Table 6.4-3). Thirteen of those were estimated in whole or in part by EJ. Heritage Apollo Program data had been used to predict some values where available and applicable.

Table 6.4-3. CPAS Input Values for PRA at IDR-2

PRA Input Value	Source
1. Drogue Confluence Failure	EJ
2. Vibration/Shock/Acceleration	EJ
3. Rigging Error	EJ
4. Far-Field Contact	Computed value, EJ
5. Near-Field Contact	EJ
6. Main Chute Entanglement	EJ
7. Drogue Chute Entanglement	EJ
8. Auxiliary/Auxiliary Chute Entanglement	EJ
9. Skipped Reefing State Failure	EJ, predicted from Apollo Program data
10. Main Chute Deployment Failure	Predicted from Mercury/Gemini/Apollo and other Missions
11. Auxiliary Chute Deployment Failure	EJ
12. Drogue Chute Deployment Failure	Predicted from Mercury/Gemini/Apollo Missions
13. Pilot Chute Deployment Failure	EJ
14. Lead Lag Failure	EJ
15. Inflation Failure	EJ
16. Reefing Cutter Fails to Dis-reef Failure	Predicted from Apollo Program data
17. Structural Failure	Predicted from material properties

At the IDR-2, reliability values for the 17 input values were presented for the cases of reentry, pad abort, and ascent abort (see Table 6.4-4).

Table 6.4-4. CPAS PRA Input Value Failure Estimates at IDR-2

	Nominal Case	Pad Abort Case	Ascent Abort Case
1. Drogue Confluence Failure	1:1,000,000	1:1,000,000	1:1,000,000
2. Vibration/Shock/ Acceleration	1:1,000,000	1:5,000	1:5,000
3. Rigging Error	1:100,000	1:100,000	1:100,000
4. Far-Field Contact	1:10,000,000+	1:10,000,000+	1:10,000,000+
5. Near-Field Contact	1:1,000,000	1:2,035	1:2,035
6. Main Chute Entanglement	1:1,000,000	1:250,000	1:250,000
7. Drogue Chute Entanglement	1:1,000,000	1:250,000	1:250,000
8. Auxiliary/Auxiliary Chute Entanglement	1:1,000,000	1:250,000	1:250,000
9. Skipped Reefing State Failure	1:1,427,000	1:1,427,000	1:427,000
10. Main Chute Deployment Failure	1:762	1:762	1:762
11. Auxiliary Chute Deployment Failure	1:1,000,000	1:1,000,000	1:1,000,000
12. Drogue Chute Deployment Failure	1:581	1:581	1:581
13. Pilot Chute Deployment Failure	1:1,000,000	1:1,000,000	1:1,000,000
14. Lead Lag Failure	1:100,000	1:100,000	1:100,000
15. Inflation Failure	1:1,000,000	1:1,000,000	1:1,000,000
16. Reefing Cutter Fails to Dis-reef Failure	1:1,523,030	1:1,523,030	1:1,523,030
17. Structural Failure	1:1,000,000	1:1,000,000	1:1,000,000

The top-level estimated reliability for the cases of reentry, pad abort, or ascent abort was also presented at IDR-2 (see Table 6.4-5). The estimates were less than the objective CPAS LOC target of 1:2,500.

Table 6.4-5. CPAS PRA at IDR-2

	Mean	95% Confidence
Nominal Reentry	1:9,502	1:2,397
Pad Abort	1:6,045	1:2,016
Ascent Abort	1:6,097	1:2,065

During RP-2, the Project conducted an EE exercise to attempt to improve the failure and uncertainty estimates for their input values. Two of the NESC parachute design and development subteam members participated as individual experts (not as representatives of the chartered assessment team). Their participation involved writing answers to a Project-developed questionnaire.

6.4.3 Review Period 3—September 2009 through April 2010

During RP-3, the target for LOC for CPAS remained 1:2,500. The target for the LRS was 1:850 for the then-current Orion design variant 606G. Recall that the LRS included CPAS plus drogue and pilot mortars, riser cutters, the fairlead fitting, NSIs, auxiliary parachute guns (if included in the design), and the CMUS.

The Project provided copies of the FTA, HA, FMEA, and PRA to the assessment team at IDR-3. The HA and FMEA had been updated significantly. The latter now listed the 'Crit-1' main riser torque-reducing keeper as a candidate design for minimum risk (DFMR) component.

The CPAS Reliability and Maintainability Report was shown to describe the Project's reliability determination methods as described previously at IDR-2:

- Reliability Estimations: Expert opinion and EJ were used to determine reliability estimates.
- Reliability Predictions: Based on historical or vendor data; examines previous reliability data collected on other Projects in order to predict component/system reliability for the activity under examination.

The Project reported that based on reliability predictions from previous generation vehicles and expert opinion and the EJ of Project personnel and AS personnel, the top LOC estimate for nominal entry was 1:9,494. This was consistent with the mean value reported at IDR-2.

The Project provided results from its recently completed EE exercise at IDR-3, but these data had not yet been incorporated into PRA input value estimates. Five experts had participated. Based on their answers the top system-level CPAS failure probability was estimated to be 1:1040. Failure to deploy a main parachute (PRA input value #10) was estimated to be 1:613 for the nominal reentry case. [Ref. 9: *Expert Elicitation Summary Report for the Crew Exploration Vehicle (CEV) Parachute Assembly System (CPAS)*, JSC 64967, ESCG-6110-09-SS-DOC-0744.]

A large variance was reported in the component reliability values. The Project reported that the experts would be asked to reexamine the resulting estimates as a sanity check.

When the CPAS architecture changes based on IDAT recommendations (described in Section 6.3.3) were adopted in November 2009, the assessment team discussed these and assessed their impact on system reliability. Near the end of RP-3, the Project indicated an intention to conduct a 'delta' EE exercise with a revised list of PRA input values that reflected the latest architecture.

6.5 Requirements, Testing, Analyses, and Verification Planning

The evolution and refinement of the Project's requirements, testing, analyses, and verification planning are described in the following sections.

6.5.1 Review Period 1—December 2008 through March 2009

At IDR-1, the draft requirements document, the PTRS [Ref. 7, Baseline version] was provided to the assessment team for review. A companion document, referred to as the *Assumptions Document* [Ref. 8, Rev. A] was also provided. The key driving requirement for CPAS was to decelerate the CM to a safe vertical descent rate for landing. The first reviewed draft of the PTRS allowed for one failed main parachute and one failed drogue parachute for cases of reentry, pad abort or ascent abort.

During RP-1, three Gen-1 test reports were reviewed and discussed. Information about the design development contractor's analysis tool, DCLDYN, was also reviewed and discussed.

For reentry and abort cases, the assessment team created a first order event tree in spreadsheet form to facilitate team understanding and to identify critical events and risks.

6.5.2 Review Period 2—April 2009 through August 2009

The PTRS (Rev. A) was released in May 2009 and was discussed at IDR-2 [Ref. 7]. A key requirement change was made for the cases of pad or ascent aborts up to a to-be-determined altitude. For aborts, CPAS would no longer be designed for 'one failed main *and* one failed drogue,' but rather for 'one failed main *or* one failed drogue.' For nominal reentry, the requirement remained 'one failed main *and* one failed drogue.'

The overall traceability for CPAS requirements was shown during RP-2 (see Figure 6.5-1). The Project had begun to integrate its PTRS with LRS requirements.

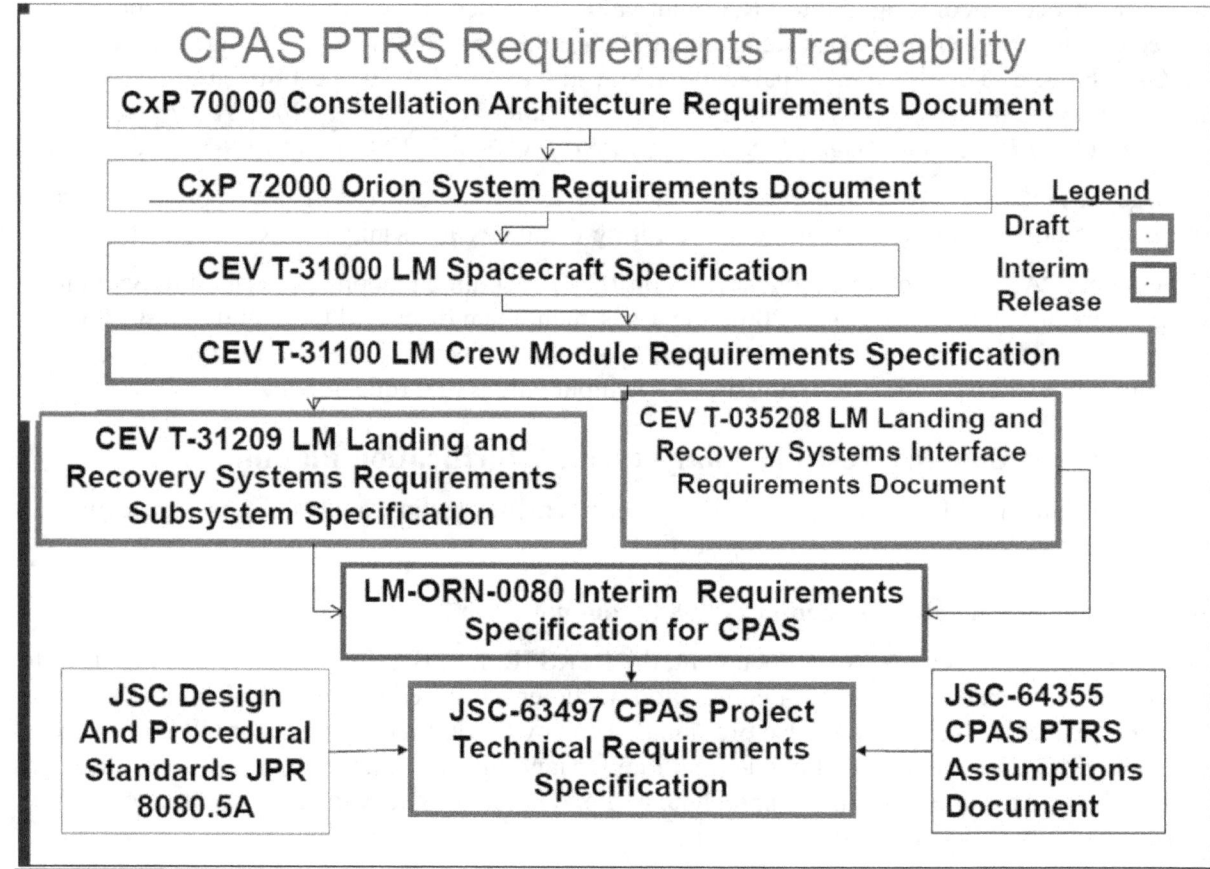

Figure 6.5-1. Requirements Traceability

The first draft of the development test matrix was provided to the assessment team during RP-2. The draft V&V document was also provided. After returning from its planned assessment hiatus, the assessment team focused much of its time during RP-2 discussing these items.

6.5.3 Review Period 3—September 2009 through April 2010

At IDR-3, the Project only cursorily reviewed the CPAS baseline design, recognizing that the ongoing IDAT activity would change the architecture later in the year.

At the review, the Project described the methods by which the draft development test matrix had been created. Test objectives had been derived principally from risks identified by the Project. These objectives then were considered in determining ground or flight testing needs. Test configurations then had been created from parameters required to create test conditions (e.g., dynamic pressure, altitude, wing loading, wake, whether a representative parachute bay or FBC was required, etc). Unique test configurations were identified, test objectives were grouped, and notional test vehicles assigned. The result was fourteen unique flight tests and two sub-scale series tests. These included main parachute performance evaluations of suspension line ratios, riser lengths, over-inflation control lines, and canopy porosity. The sub-scale drops were to evaluate the uncertainties related to CPAS separation from the parachute test vehicle (PTV).

The assessment team had assessed the test matrix independently and identified much of this detail prior to the IDR-3 (see Figure 6.5-2). The assessment team presented its second interim outbrief to the Project the same week in September 2009 as the IDR-3.

When the CPAS architecture changes based on IDAT recommendations were adopted in November 2009, the assessment team discussed these and created a revised 'event spreadsheet' for this architecture to help the team identify areas of risk and concern that would necessitate test plan changes.

Risk/Objective Number	Motivation/Risk Level	Name	Motivation/Risk Description	Objective	Dependence	Ground or Flight Test
14	Risk/Low	Main Riser Twist	Main chute deployment with riser twist	Observe effects of twisted main risers on main deployment	Generic	Flight
15	Risk/High	Main Overload	Main Overload Test	Verify parachutes can withstand conditions that cause loads 20-30% over Design Limit Loads	Generic	Flight
16	Risk/Medium	Main Riser Abrasion	Abrasion between main risers	Twist Test results will determine what flight test is required	Generic	Ground Primary
New 3 (36)	Risk/High	Torque Limiter	Test for torque limiter performance	Torque limiter performance with one main chute in bay. Test needed for model validation.		Ground
18	Risk/Medium	Drogue Skipped Stage	Drogue -- Inaccurate skipped stage modeling leading to improper design of reefing schedule	Verfiy loads on gusset. Provide data to improve skipped stage modeling (fill time, Ck, CdS(t) etc.)	Generic	Flight
19	Risk/Medium	Main Lead/Lag	Main -- Inaccurate lead/lag modeling leading to improper design of reefing schedule	Demonstrate Load sharing assumptions /load profile assumptions (skipped stage clusters)	Both	Flight
20	Risk/Medium	Main Skipped Stage	Main -- Inaccurate skipped stage modeling leading to improper design of reefing schedule	Improve Skipped stage modeling (fill time, Ck, CdS(t) etc.)	Generic	Flight
21	Risk/Low	One Drogue Two Main	1 drogue failure, 1 main failure	Evaluate 1 drogue, 2 main nominal case to determine if cluster to limit load may not achieve 65/35.	Generic	Flight

Figure 6.5-2. Representative Portion of Development Test Matrix

6.5.4 Analysis

Specific documents provided from attending these meetings are detailed in the following sections.

The Project conducted three IDRs during the assessment team charter. The presentations that were reviewed for technical content from the IDRs are System_IDR_12-08-08.ppt, System_IDR_II_5-20.ppt, System IDR II 9-21-09_Primary Charts.ppt. Each presentation details performance results and assumptions used in designing CPAS parachutes and hardware. The presentations describe the primary CPAS simulation tool, DCLDYN, and the models used to characterize the performance of the parachute system. An overview of multiple aspects of CPAS are presented, including the Con Ops, materials, interfaces, environment, proposed chute difficulties (torque limiter, confluence fitting, bag retention system, etc.), mass properties, safety, and reliability.

During the IDR-3, a breakout session on simulation models and methods was conducted. Presentation material was detailed in CPAS_Sim_Methods_2009-09-21, Osiris_HiFiModel_Overview.ppt, and PA1_CPAS_GNC_TIM.ppt. The session presentations describe analysis and reconstruction approaches of parachute test data by modeling and simulation. The session also helped to identify how one of the primary simulations for Orion (Outdoor Scene and Infrared Image Simulation or OSIrIS) would incorporate its own parachute model with comparisons to models in Decelerator System Simulation (DSS).

SE&I Integrated Product Team (IPT) teleconferences were attended by various assessment team members from January 2009 through March 2010. The meetings were reviewed by the assessment team for CPAS design decisions and methodology. Hardware IPT and CPAS MVP planning meetings were also attended during the same time frame.

IDAT charts from the period of August 2009 to October 2009 were reviewed for design decisions and methodology. Specific areas of NESC interest in relation to IDAT are the five chartered trade spaces:

1. Parachute Architecture—2 Drogue/3 Main, 3 Drogue/3 Main, 2 Drogue/4 Main
2. FBC Removal—Segmented versus Monolothic FBC
3. Forward Bay Volume—Backshell angle modification, FBC modification, structural design changes
4. LAS/CM Forward Attach—CM/LAS R&R design mitigation, LAS/CM attachment blisters/wells
5. LAS Architecture—Orientation for post-abort LAS jettison, thrusters/jettison motor/tip chute jettison assist, straight-to-mains

Drop test reports from the CPAS Development Test Gen-1 Series for the pilots, drogues, and mains (ESCG-4390-080-SP-DOC-0029, ESCG-4390-08-DOC-0028, and ESC-4390-08-SP-DOC-0036) were reviewed for simulation model accuracy and prediction of parachute performance.

The Apollo Program Experience Report (NASA-TN-D7437) and the Apollo 15 Parachute Failure Report (NASA-TM-X-67439) were reviewed for applicable lessons learned.

Various PowerPoint®, Word®, and Excel®, documents provided from CPAS were reviewed that include the CPAS organizational charts, MVP, PTRS, assumptions document, interface documents, test plans, and reliability PRA.

Various papers were reviewed detailing the Orion Guidance, Navigation, and Control (GN&C) architecture utilized by CEV to guide the entry vehicle to the parachute deploy conditions and the associated triggers used to deploy the parachute. AAS 07-071 "NASA CEV Reference GN&C Architecture," American Institute of Aeronautics and Astronautics, "Orion Entry, Descent, & Landing Simulation," and "GN&C Design and Data Book, Volumes I, IIa, and V."

6.5.5 Master Verification Plan

An introduction to the Project's iterative verification strategy was provided at IDR-3 (see Figure 6.5-3).

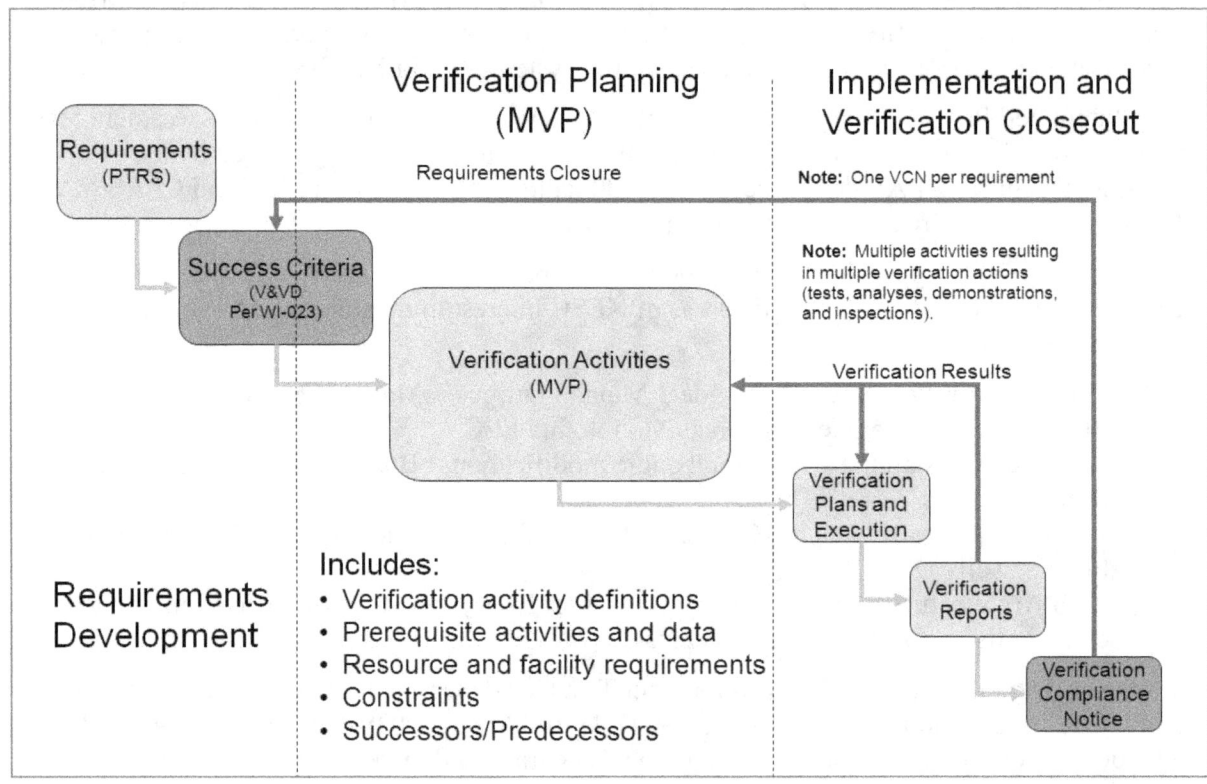

Figure 6.5-3. Verification Strategy at IDR-3

The Project conducted an 'MVP kickoff' meeting in October 2009 and provided the first draft of its MVP. Four types of verification activities (VAs) were described in the MVP: verification by analyses, testing, inspection, and demonstration. The VAs were frequently described figuratively as being identified at the 'atomic' level. (See Table 6.5-1.)

Table 6.5-1. MVP Verification Strategy

Verification by Analysis
High-fidelity dynamic/structural simulation process for Monte Carlo analysis to provide static and dynamic loads including protuberance interaction and friction and vehicle dynamics response to parachute and harnessing effects.
Likely analogous to landing loads analysis process.
Development team to develop, verify, validate, and accredit simulation.
Analysis team to derive, analyze, and verify structural margins at all points.
Requires Simulation tool for dynamics of adverse attitude riser contact.
Requires Additional finite element model analyses (stress and loads/dynamics).
Verification by Test
Series of tests to validate/anchor simulation models (parachute effects, aero effects). • Spin tunnel testing (new, cheap—around $50K) • ¼-scale drop testing (new, in CPAS impact estimates for 3d/3m) • Full-scale pallet testing (new, in CPAS impact estimates for 3d/3m) • Pad Abort (PA)-1 flight test • CPAS PTV drop testing • Ascent Abort-2 flight test • PA-2 flight test
Static ground testing using structural test authority for off-nominal loads applications including cases when wrapped around heatshield with opportunity to limit these as flight simulations are matured.
Verification by Inspection
Inspection (and perhaps some demonstration) for riser/structure interaction late in assembly sequence to verify the assessment team captured all possible interference.
Verification by Demonstration
Utilize current CPAS drop testing for system certification.
Development PTV tests: • 1 with drogue failure (To Be Decided (TBD) attitude/rate behavior)
Qual PTV tests: • 1 off-nominal attitude (e.g., apex forward) • 1 with drogue failure (TBD attitude/rate behavior)

At the MVP kickoff, the Project discussed plans to conduct follow-on activities to consolidate and aggregate VAs and to extend the tasks to accommodate hazard-control requirements, interface controls and other requirements, and qualification risks. They stated an intent to comprehensively capture VAs during the engineering development phase.

After the MVP kickoff and during subsequent Project meetings, the assessment team uncovered insufficient Project understanding of the test and verification recommendations the assessment team had provided in September 2009. To remedy this, the assessment team conducted a question-and-answer meeting in December 2009. This document is included in Appendix C.

6.6 Organizational Complexity

The assessment team examined the organizational framework in which the Project was operating and discussed risks it could pose to the design and development of a reliable system. (See Figures 6.6-1, 6.6-2, 6.6-3, and 6.6-4.)

It is pertinent to describe the partition of responsibility for development of Orion hardware here. The CM included the gussets, tunnel, bulkhead, FBC, avionics, riser cutters, and LRS. The LRS was comprised of two principal components: the Aerodynamic Decelerator System (ADS) and the CMUS. The ADS was comprised principally of CPAS, plus other subcomponents including mortars, riser cutters, and auxiliary parachute guns (if any). (See Figure 6.6-5.)

AS was contracted through Jacobs ESCG to provide design and development data, test hardware, and test data for the Project. AS did not have a production contract for CPAS at the time of this review.

The Project was responsible for parachutes and their components, risers, deployment bags, confluence fittings (if any), torque reducer (if any), and other ancillary components.

The Orion Project's lead contractor, LM, was responsible for several other components critical to CPAS reliability, including all mortars, all riser cutters, the fairlead fitting, and auxiliary parachute guns (if any). LM was also responsible for integrating CPAS into the larger LRS, which included the CMUS. LM also was responsible for integrating the LRS into the CM and for CM features critical for CPAS reliability, including the FBC and its jettison system, and the forward bay gussets and tunnel.

The assessment team's understanding was that the design development data and test data were to be provided by AS (through its contract with Jacobs ESCG) to the Project. The parachute system ultimately would be provided to LM as GFE.

The CPAS SE&I (see Figure 6.6-1) and S&MA (see Figure 6.6-2) functions were shared by NASA and Jacobs ESCG, but top-level integration authority resided with the LM Engineering Review Board (see Figure 6.6-4).

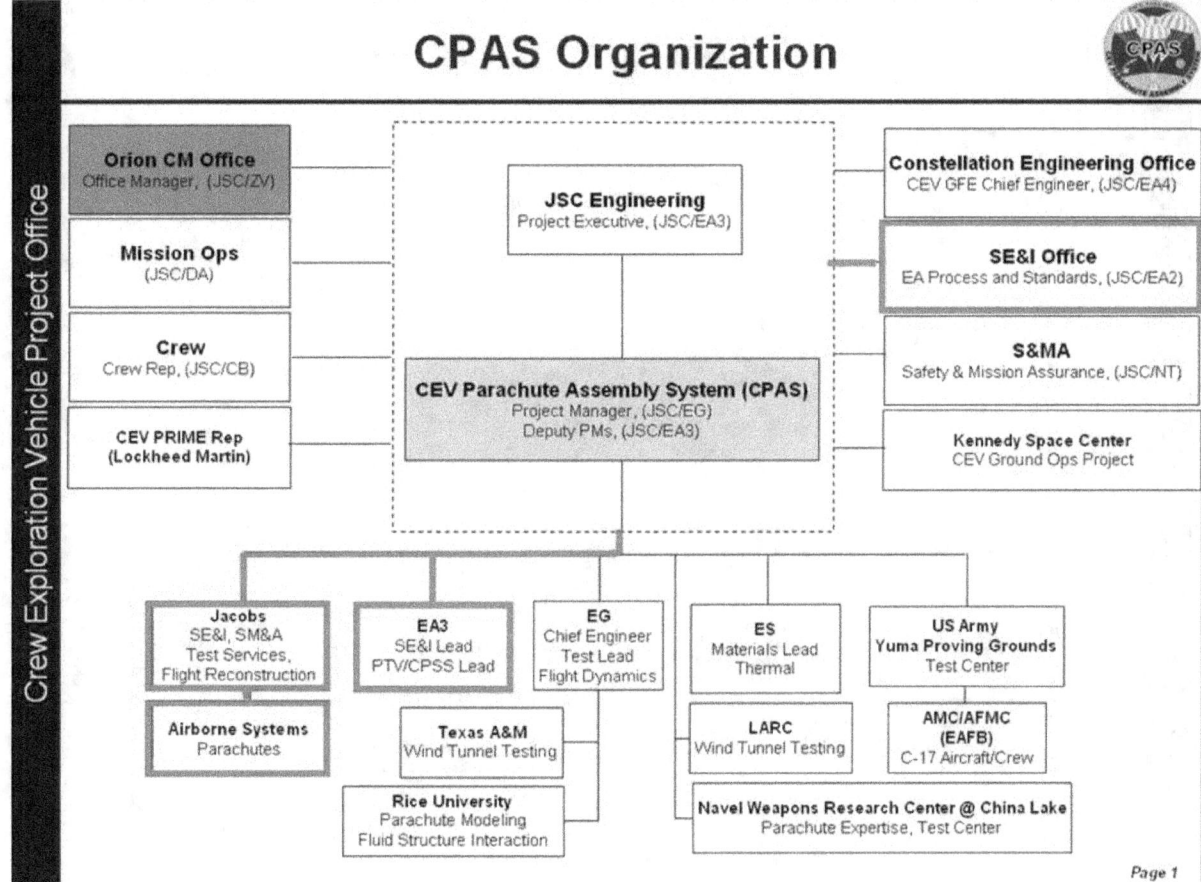

Figure 6.6-1. CPAS Organization, SE&I

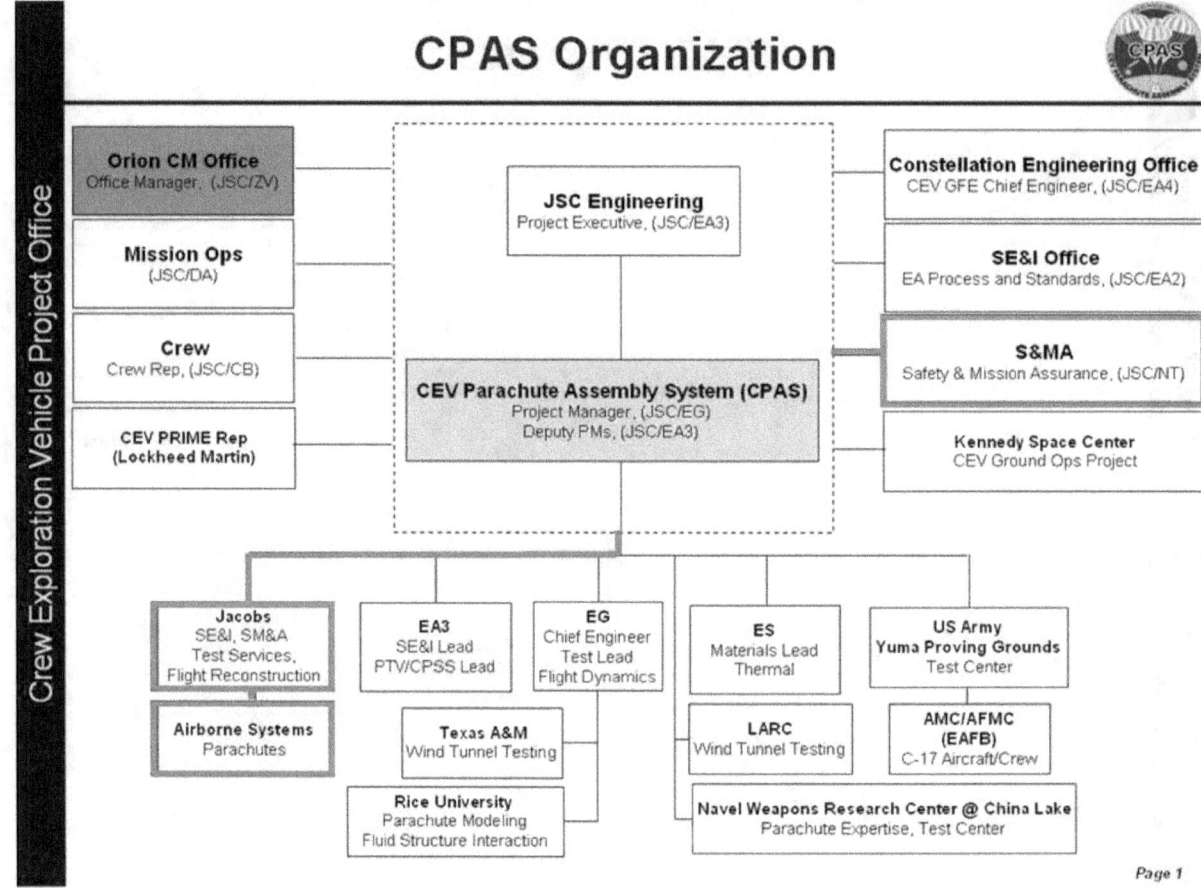

Figure 6.6-2. CPAS Organization, S&MA

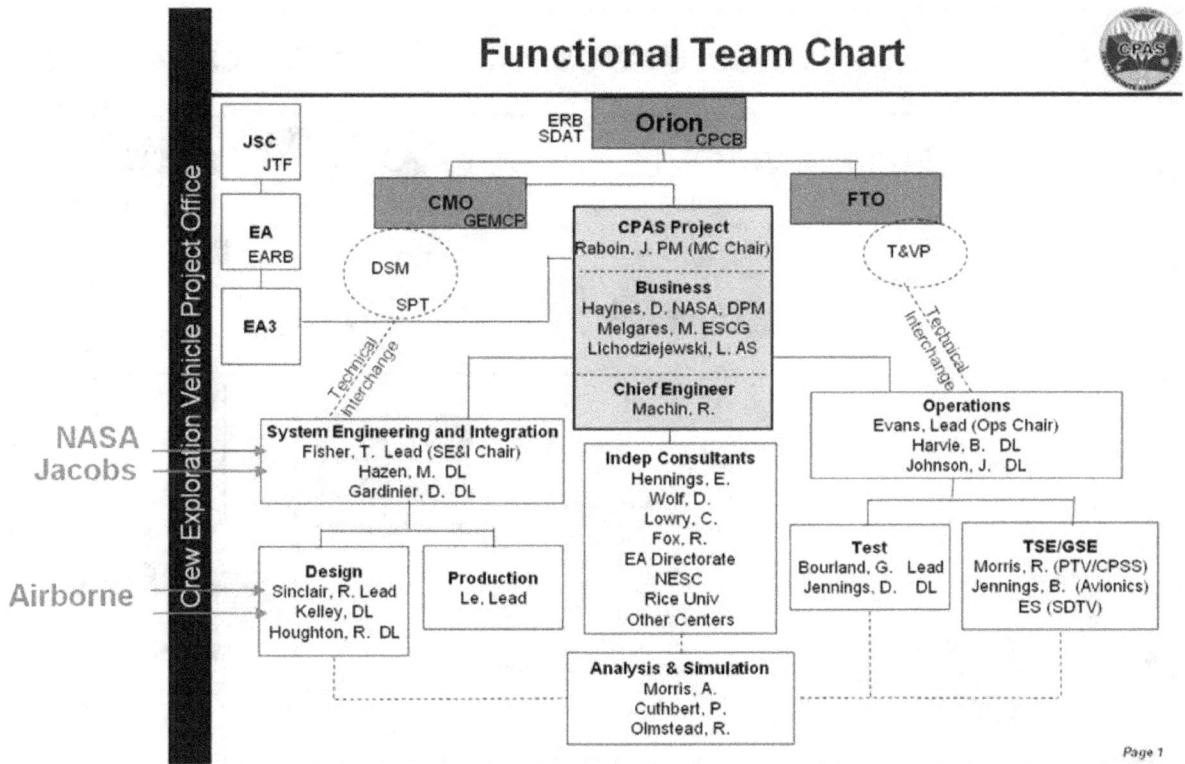

Figure 6.6-3. *CPAS Project Functional Team Organization Chart*

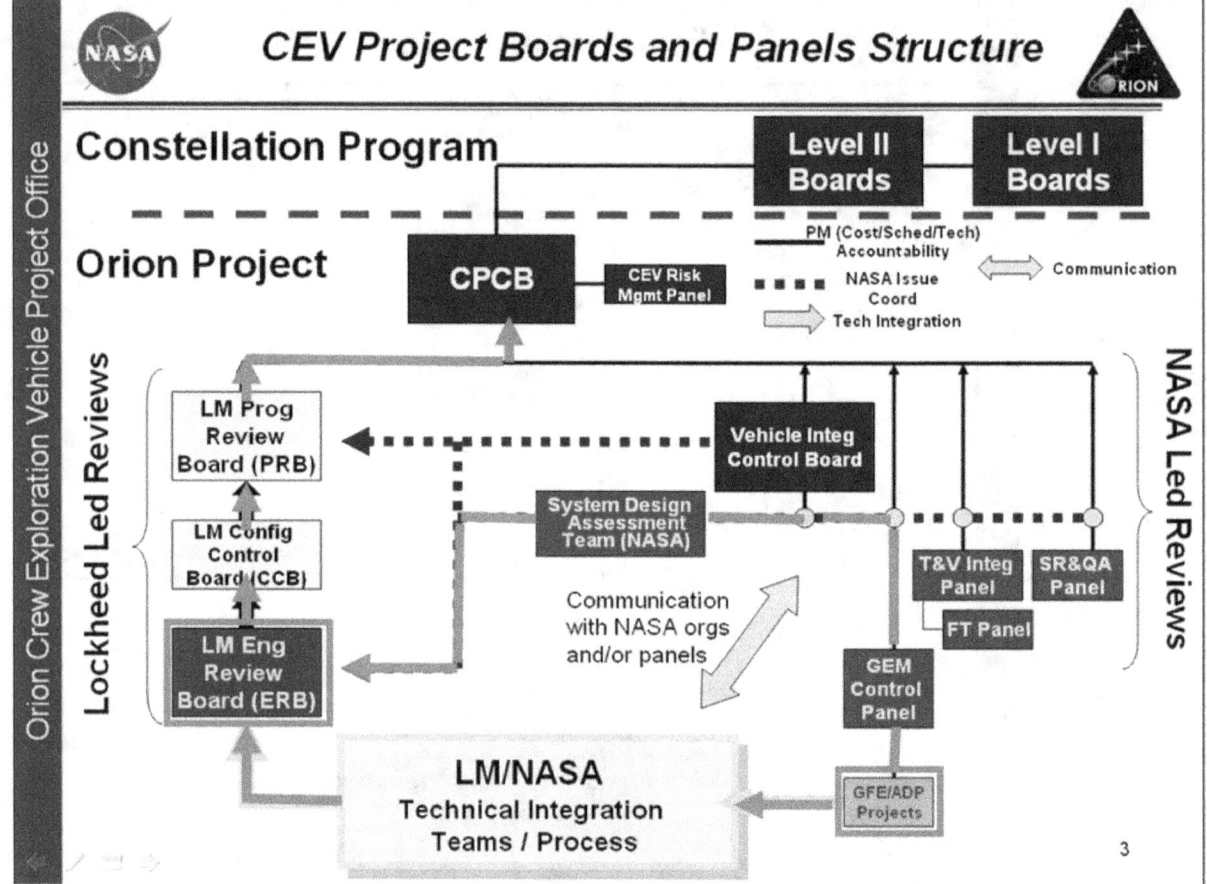

Figure 6.6-4. GFE CPAS Boards and Panels Review Paths

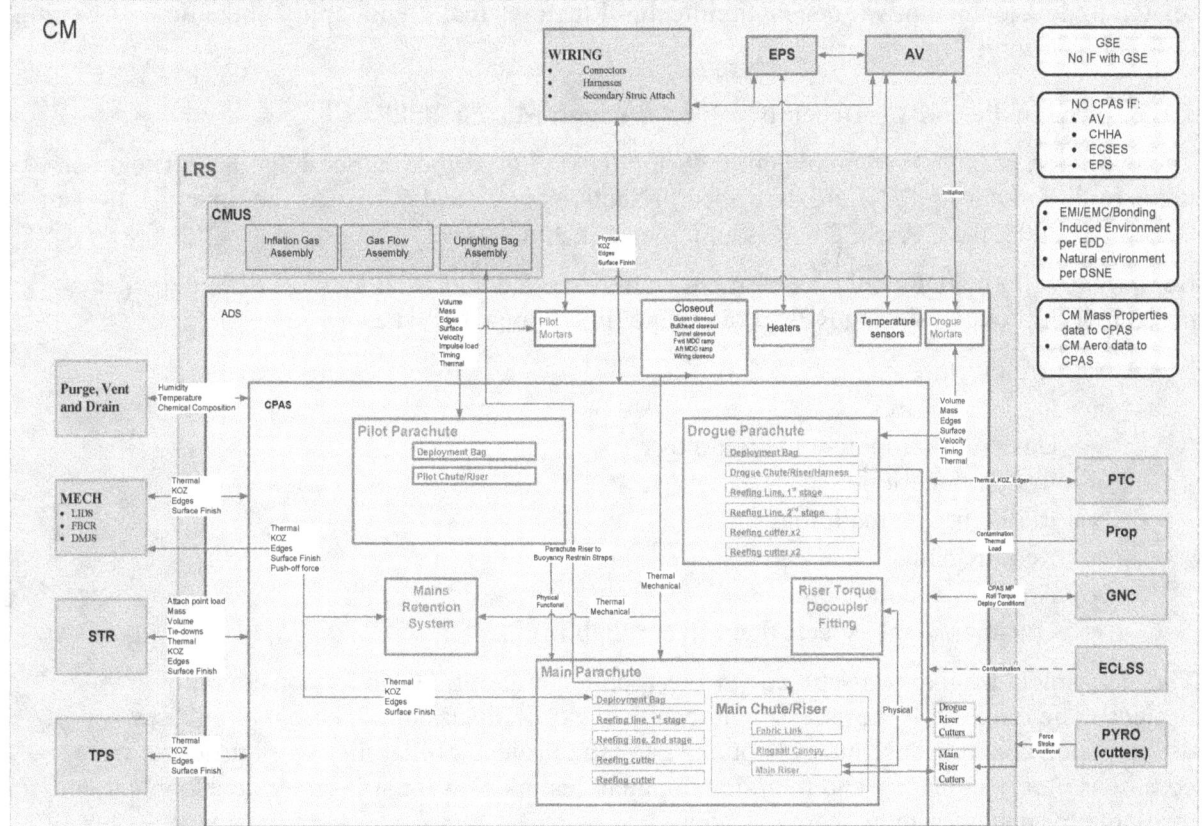

Figure 6.6-5. CPAS/ADS/LRS/CM Interface Definitions at Conclusion of RP-3

7.0 Data Analysis

This section summarizes the assessment team's analyses of relevant Project information described in Section 6.0. Section 7.1 provides an analytical context for the three sets of findings, observations, and *interim* NESC recommendations that are reproduced in Appendices B, C, and D. Section 7.2 provides team analyses supporting the findings, observations, and *final* NESC recommendations provided in Section 8.0.

7.1 Interim Analytical Assessment of Project Information

The three NESC stakeholder summaries provided to the Project in April 2009, September 2009, and April 2010 provided time-sensitive *interim* recommendations intended to lower project technical and schedule risks. The CPAS architecture changed considerably during this time, and the assessment team's interim deliverables reflected those changes as they occurred. Many related to developmentally transient architectural features that do not apply at the time of writing of this final report.

7.1.1 Review Period 1—December 2008 through March 2009

The evolution of the CPAS architecture during RP-1 is described in Section 6.3.1. The other items reviewed during RP-1 are described in Sections 6.4.1 and 6.5.1. At the time of IDR-1 in December 2008, the CPAS PDR was anticipated in June 2009.

Beginning with the IDR-1, the assessment team reviewed a large amount of design data and project work products during RP-1, covering a broad range of issues:

- Architecture
- Design features
- Design development methodology
- System function
- Reliability methodology
- Requirements
- Analysis and modeling
- Organizational aspects that affect reliability

The assessment team also discussed various architectural options under consideration by the Project. The request for the assessment team had explicitly assumed there was an "existing" design, but a concrete baseline did not exist and the Project's reliability work products were commensurately immature and their development test plans were still under development and not available for the assessment team to review.

To compile and organize the many relevant issues discussed by the assessment team, a preliminary list of 336 factual statements (FSs) and 96 experience-based observations (EOs) was created.

The assessment team derived from this list 20 findings, 24 observations, and 30 *interim* NESC recommendations. These were approved by the NRB and outbriefed to Project stakeholders at AS in Santa Ana, CA, in April 2009.

The first stakeholder outbrief and the supporting list of FSs and EOs can be found in Appendix B.

The *interim* NESC recommendation R-3 from the first stakeholder outbrief is discussed further in Section 7.2.1.1, as part of narrative that supports the *final* NESC recommendation R-1 in this report.

7.1.2 Review Period 2—April 2009 through August 2009

The evolution of the CPAS architecture during RP-2 is described in Section 6.3.2. The other items reviewed during RP-2 are described in Sections 6.4.2 and 6.5.2. During RP-2 the CPAS architecture continued to evolve. At the time of IDR-2 in May 2009, the CPAS PDR was anticipated in November 2009.

After discussing issues learned at the IDR-2, the assessment team took a 2-month hiatus, discontinued monitoring CPAS meetings and discontinued conducting weekly team meetings, until July 2009. This stand-down was to allow the Project to make progress toward a more concrete design architecture definition, corresponding reliability work products, and a draft development test plan. The assessment lead attended the Orion Subsystem Design Review in June 2009 and the System Module Review in July 2009, to remain abreast of CPAS design developments.

The assessment team began meeting again in July 2009 when a draft development test matrix and a draft verification plan were made available by the Project for review. While keeping abreast of design architecture options being vetted by the IDAT, the assessment team focused most of its discussions on the Project's development test plans, analyses plans, and verification activities. The assessment team discussed test matrices derived from functional decomposition of systems and created a sample *Functional* Verification Matrix. The assessment team also discussed applications of experimental design in planning CPAS development and verification activity.

The assessment team created a second outbrief of three findings, two observations, and two *interim* NESC recommendations. These were approved by the NRB and outbriefed to Project stakeholders at JSC in September 2009.

One of the two recommendations provided specific guidance on the application of DOE techniques that can improve the effectiveness and resource-efficiency of the Project's development testing effort. The other recommendation outlined a process for creating a comprehensive matrix of candidate verification tasks that the Project could use as a check against their nascent verification plan, to identify potential gaps and/or redundancies.

The assessment team included supplemental material in the stakeholder outbrief, including an introductory explanation of potential benefits of using methods of statistical DOE, in particular comparison to using 'one factor at a time' testing methods, and the sample Functional Verification Matrix. During RP-3 the assessment team created a list of frequently asked questions (FAQs) pertaining to DOE (described in the following section).

The second stakeholder outbrief, the sample of a Functional Verification Matrix, the DOE tutorial, and the list of FAQs can be found in Appendix C.

The *interim* NESC recommendation R-31 from the stakeholder interim outbrief 2 is discussed further in Section 7.2.3.2 as part of narrative that supports the *final* NESC recommendation R-6 in this report.

7.1.3 Review Period 3—September 2009 through April 2010.

The evolution of the CPAS architecture during RP-3 is described in Section 6.3.3. The other items reviewed during RP-3 are described in Sections 6.4.3 and 6.5.3. At the time of IDR-3 in September 2009, the CPAS PDR was anticipated in March 2010. By the end of RP-3, the PDR was scheduled for August 2010.

At the beginning of RP-3, the assessment team composition was adjusted. The assessment team's structures discipline expert and one parachute design development expert left the team. A second DOE expert and a heritage systems safety expert were added to the team.

The assessment team continued to discuss various design options being vetted by the IDAT.

The assessment team also continued to discuss the Project's draft verification plan. The assessment team detected that Project personnel may not have fully understood the *interim* NESC recommendations that were communicated in September 2009, so the team compiled a list of clarifying answers to questions that were being asked by CPAS personnel. This document of FAQs was not an NRB-approved product. The assessment team conducted a meeting with the Project in December 2009 to discuss these questions.

After the Project adopted IDAT recommended architecture, the assessment team assessed the impact to design reliability.

The assessment team created a third outbrief of 33 findings, 20 observations, and 23 *interim* NESC recommendations. These were approved by the NRB and outbriefed to Project stakeholders at JSC in April 2010.

The assessment team put backup material in the outbrief including an example of reliability growth trending analysis and an example of the potential problems with one-factor-at-a-time testing.

The stakeholder outbrief 3 and the two noted example items can be found in Appendix D.

7.2 Final Analytical Assessment of CPAS Project Information

This section provides a narrative summary of the team's assessment of Project information current at the end of the overall assessment period, in April 2010.

Throughout the entire period of assessment review, the CPAS architecture was evolving. During the third review period, a final architecture was selected based on recommendations from an IDAT chartered to resolve key CPAS integration issues. Several outstanding *interim* NESC recommendations were addressed by this new architecture.

Included in this section are team assessments resulting in findings that support the *final* NESC recommendations in Section 8.0. These recommendations were approved by the NRB prior to completion of the report and provided to Project stakeholders prior to the CPAS PDR in August 2010.

7.2.1 Architecture

7.2.1.1 Pack Volume Risks

The volumetric requirements for any parachute system are driven by vehicle recovery weight, maximum deployment dynamic pressure, allowable packing density, materials used, and terminal descent velocity. Higher vehicle weight or higher dynamic pressure would demand stronger and possibly larger parachutes, which in turn would demand more volume. Once a parachute compartment volume has been allocated, it can be difficult to change the design to provide more volume in response to future demands.

The CPAS Subsystem's PTRS specified that the CM was required to provide volume to ensure CPAS main parachute density was ≤ 38 lbm/ft³. For the first two review periods, the projected main parachute pack density was between 43.7 and 54.2 lbm/ft³.

There was consensus on the assessment team and across the wider parachute community that as parachute pack density increases, reliability risks also increase. Risks include damage to hardware items and soft materials. Damage to some parachute components can lead to parachute inflation malfunctions that cause loads to exceed design limit loads. Damage to components can also cause them to fail at loads below design limit loads. Either of these outcomes can result in catastrophic parachute system failures.

Some parachute systems on emergency escape systems or unmanned weapons systems have successfully incorporated high-density parachute packs but these were not primary landing systems used on manned vehicles.

During the Apollo Program ELS Block II development program, parachute performance requirements increased but the volume allocated to the main parachutes remained constant, resulting in increased packed density. On the Apollo Program, increased rates of parachute component damage were observed in proportion to increases in pack density. Innovative packing solutions with rigorous certification efforts and personnel training requirements were needed to mitigate these risks. Functional verification could not be performed on these high-density parachute systems during much of their manufacture, so rigorous, time-consuming, and costly manufacturing and installation practices were implemented.

The Project should create and maintain a budgeting process for volume and pack density. The volume assigned to CPAS in the CM forward bay should be cataloged as to gross dimensions, volume, shape, and draft. This should include the space for all CPAS components, including but not limited to the main parachutes, drogues, pilot chutes, risers, bridles and associated hardware, and any other systems stored in this volume. A volume margin should be added and each design change proposed during development should be evaluated for impact to this budget. The assessment team observed that virtually every escape or recovery parachute system experienced increases in weight requirements during development, including the Apollo Program ELS. This history suggests that Orion CM weights will likely increase once the vehicle enters operation, making it necessary to continue to monitor the vehicle parachute compartment volumetric margins.

The volume provided by the new CM OML described in Section 6.3.3 allowed the CPAS main pack shape to be simplified and the pack density to be reduced to below the required level. However, the assessment team assessed that the continued risk of future volume demands was improperly prioritized by the Project.

F-1. The CPAS Project has not implemented a formal process to manage volume growth of components in the CM forward bay during design, development, evaluation, and operation.

7.2.1.2 Roll Control Risks

To reduce water-landing loads on the CM structure and its crew, the CM was to be oriented to a toe-forward orientation just prior to water landing. To facilitate this, CPAS included a metallic torque reducer keeper that would deploy with the main parachutes and gather the main risers to a point above their single point attachment location on the CM. This would allow a section of gathered risers to 'twist up' and 'untwist' with less than 450 ft·lbf when the CM was suspended under its opened main parachutes. With the roll torque limited as such, the CM RCS would have sufficient authority to roll the CM to an optimum water entry angle prior to landing.

The assessment team assessed that deployment of a keeper on the CPAS main risers had no apparent design precedent and posed risks to CPAS reliability. Extensive testing would be necessary to quantify these risks and demonstrate reliable function of this DFMR component.

The assessment team thought that using the RCS while suspended on parachutes could also pose a risk of causing burning or oxidizing damage to parachute textile materials, especially at high CM angular rates and attitudes. This was the suspected cause of one main parachute failure during descent of the Apollo 15 CM.

In the course of discussing the inclusion of the keeper, the assessment team assessed other risks that may arise when using this method to achieve CM orientation relative to ground track heading. To impact a wave slope in the necessary manner would require knowing and managing a number of variables in a short period of time near the water. These variables would include the descent rate of the CM, the number of degrees of roll reorientation required, the target roll rate, the swing angle under the parachutes, the drift rate and direction, and the tolerance in the heading. Other possible variables associated with proximity to the water surface could include wind shears and gusts.

A successful reorientation maneuver may also require a sophisticated sensing system to determine water surfaces conditions at touchdown for computation of the CM's terminal orientation requirement. The assessment team knew of no precedent for this type of system. It was not within the assessment team's scope of review to investigate onboard sensor system requirements.

The assessment team concluded that the Orion Project should consider chartering an IDAT, perhaps delegated to the previous IDAT's successor, the Earth Landing System Functional Integration Team (ELS-FIT), to conduct a comprehensive assessment of the overall feasibility and ramifications of using the CM RCS to perform the function of reorientation of the CM prior to water landing.

The IDAT should include representatives from the CM subsystems that interface with or are dependent on the RCS function, including CPAS. The assessment should address the risks of inclusion of a deployable metallic keeper in CPAS and should also include an assessment of LRS verification plans.

The IDAT assessment scope could include:
- The overall feasibility and ramifications of using the RCS for touchdown roll control
- Assessment of the risks (likelihood/consequence) versus benefits to CPAS and other subsystems
- The impact of prohibiting the use of the RCS once the FBC panels are deployed
- The impact of a requirement to not dump RCS fuel in any direction where heat, fuel, or oxidizer can reach any part of the parachute system
- Inclusion of a deployable metallic torque limiter in CPAS
- Design of the CM structure for water impact at worst-case combinations of water-surface conditions and vehicle-landing attitudes
- The risks of landing with fuel in the RCS tanks
- The effects of parachute oscillation
- The ability to sense CM drift direction
- The ability to sense water-surface conditions beneath the CM prior to landing
- The methods to determine which heading to take to impact a dynamically changing wave slope just right
- RCS command sequencing
- The extent and complexity of an adequate development and certification plan for the RCS

F-2. Deployment of a torque reducer keeper on the CPAS main risers has no design precedent and poses risks to CPAS reliability. Extensive testing will be necessary to quantify these risks and demonstrate reliable functionality.

F-3. Using the CM RCS to orient the CM into a toe-forward position just prior to water landing is unprecedented and is an integrated issue that introduces risks to CPAS and the CM structure.

7.2.2 PRA and Other Reliability Products

7.2.2.1 PRA Estimation Methods

At the first IDR in December 2008, a draft of the RAESR was presented that included a HA, FMEA, and FTA. The RAESR provided a functional description of the system and decomposition based on the current architecture, which was immature at this time. In addition, a PRA was presented that estimated catastrophic main parachute failure leading to LOC at 1 in 217 without a statement of uncertainty.

The PRA estimate was dominated by a failure mode seen in the SSP SRB historical development data that involved re-contact during parachute deployment. During design evolution, the FTA and FMEA were updated in accordance with design changes. The FMEA was used to track single point failures, which were briefed to the Project design team. However, the NESC assessment team noted that reliability analyses were severely limited by a lack of appropriate test data to estimate the FTA probabilities. In the following paragraphs, significant events in the reliability and PRA's evolution are highlighted.

The assessment team had determined that the IDR-1 LOC estimate was invalid because the SSP SRB failure mode was not applicable to the CPAS architecture. Consequently, the CPAS LOC estimate was considered not accurate and with great uncertainty. The PRA also had gaps caused by a lack of valid test data on which failure estimates could be based.

As the CPAS architecture evolved toward a system similar to the Apollo Program ELS during RP-3, additional data were sought from historical files from the ELS development program to use in the PRA. At the end of the team's assessment, a significant amount of Apollo ELS data had been discovered but had not yet been integrated into the PRA.

An EE activity was performed to fill the test data gaps and replace rough order of magnitude EJ estimates. The EE was based on the architecture current at the time of the exercise. It was intended to derive a relative ranking of potential failure modes from a small group of parachute system experts.

The results of the EE were used in subsequent PRA LOC estimates. Details of the EE are provided in Reference 9.

Because of the existing FTA's complexity and the rapidly changing design, a simplified FTA was developed at the request of the CPAS SE&I lead that was more "interpretable" by design engineers to support design and development decisions. This second FT attempted to capture higher-level functional failures rather than individual failure modes. While this FT was more understandable by the Project design team, the original, more complex FT simultaneously was maintained during the assessment period and was used for "official" PRA estimates.

The Project provided results from the EE at IDR-3, but those data had not yet been incorporated into the PRA basic event estimates. Five experts had participated. Based on their answers a top system-level reliability was estimated to be 1:1,087. Failure to deploy a main parachute (PRA input value #10) was estimated to be 1:231. A large variance was reported in the component reliability values. The Project reported that the experts would be asked to reexamine the resulting estimates as a sanity check. The Project reported that inputs for PRA failure mode values were to be used to determine top system-level reliability.

The failure probability estimate data derived from the 2009 EE exercise exhibited large variance that greatly limited its applicability for generating CPAS PRA input values. The large variance was due mainly to the disagreement between the five experts involved in the study.

The failure probability estimate data derived from the 2009 expert elicitation exercise are invalid for the Gen-2 architecture. The Project is attempting to modify the EE results to account for changes in the Gen-2 architecture by removing a number of failure modes that are no longer part of this architecture. The assessment team agreed that modification of the results in this manner would not lead to valid probability estimates for the Gen-2 architecture.

The assessment team observed that the PRA appeared to be used principally for computing top-level LOC estimates to communicate the Orion Project, but it also had influence on the development test plan, consistent with the Orion Project's risk-based approach. This approach would use the PRA to define and prioritize risks and to define the resources required for testing.

The assessment team considered this approach insufficiently comprehensive for a development effort. For the Project's PRA/risk-based approach to succeed, all primary risks would have had to have been identified.

As the FMEA evolved, critical failure modes and total failure modes were significantly reduced (see Table 7.2-1).

Table 7.2-1. Evolution of FMEA from RP-1 to RP-3

FMEA December 2008:	
Criticality	Amount
1	145
IR	17
1R2	739
1R3	110
1R4	8
Total:	1019
FMEA September 2009:	
Criticality	Amount
1	2
1R	0
1R2	66
1R3	0
1R4	6
Total:	74

The reliability analyses, including the RAESR and PRA, were reviewed throughout the overall assessment period. The assessment team found the traceability and veracity of the data consistently problematic, particularly in the input data to the PRA. Estimated risks contained historical data, new test data, simulation, and EE results. These sources were neither explicitly noted in the analyses nor communicated to the Orion Program. The assessment team found inappropriate statistical distribution assumptions in both experimental and simulation data that rendered them either under- or over-conservative. The assessment team also found early risk estimates with no statements of uncertainty, which did not support risk-informed decision-making.

The original PRA LOC estimates were found to be unsubstantiated due to the use of inapplicable SSP SRB data based on re-contact during deployment due to the dissimilarity between the CPAS and SRB architecture.

The assessment team's analysis indicated that the lack of applicable test data presented a formidable challenge to obtaining adequate PRA estimates.

Consequently, the assessment team found that Apollo Program ELS data should be used for PRA inputs whenever appropriate (by similarity) and when available. During the assessment, a cache of Apollo Program reports were discovered in a storage facility managed by NG. These data will be more directly applicable to the current CPAS architecture and should improve PRA estimates.

The assessment team reviewed the design and analysis of the EE exercise and found that the results were highly variable among the experts. The methods employed appeared to assume that the elicited events were independent, whereas some were dependent. The assessment team found that the experts were asked to compare events at disparate levels. The large variability between the experts suggested lack of agreement and indicated problems with the elicitation protocol, including the events collection, their precise definition, and/or the analysis approach.

The EE was based on an obsolete architecture. The team assessed that deleting PRA failure events to attempt to model the current architecture would render the EE ranking invalid. If another EE were to be conducted, a structured methodology such as an Analytical Hierarchy Process or a similar decision-ranking technique should be employed. If another EE were to be conducted, the Project would benefit by seeking an EE discipline expert's assistance.

F-4.   Failure probability estimate data derived from the 2009 EE exercise exhibited a large variance, greatly limiting its usefulness for generating CPAS PRA input values.

F-5.   Failure probability estimate data derived from the 2009 EE exercise are invalid for the Gen-2 architecture.

O-1.   A large cache of Apollo Program ELS test and reliability reports has been located and is being retrieved by the Project. Some of these data will be applicable for generating CPAS PRA input values because the ELS and the CPAS Gen-2 architectures are functionally similar.

O-2.   Top-level system failure probability estimates often are estimated or predicted prior to PDR, but tests and analyses ultimately must be performed to validate and reduce PRA input value uncertainties.

7.2.2.2 Use of PRA Estimates for Design Decisions

The impetus for initiating this NESC assessment was an unacceptably high LOC probabilistic risk estimate reported to the Orion Project. This estimate, based erroneously on inapplicable SSP SRB data, was conveyed to CxP and CEV Project decision makers without a statement of uncertainty. If the uncertainty regarding the data's inapplicability and lack of statistical confidence in the LOC point estimate had been communicated, more actionable information would have been provided to these decision-makers. Proper communication of uncertainty, particularly for mission-critical parameters such as LOC, is imperative.

The methods used by the Project to generate the PRA input value failure probability estimates evolved considerably throughout the assessment. Supporting data sources can include heritage test data or operational data, original test data, EJ, and EE.

PRA is best utilized for optimizing the system's reliability. Instead of simply generating a LOC number, PRA can provide a beneficial tool for improving system reliability. PRA tasks are best completed with the intention and motivation for improving a system's reliability.

The assessment team concluded that the Project should convey to top-level design decision makers that there presently are large uncertainties associated with the PRA estimates, caused by a complex and unique parachute design coupled with minimal field test data. For PRA estimates, data sources (e.g., historical records, CPAS test, analysis, expert opinion), accuracy limitations, and relevance to the CPAS design must be communicated.

O-3. The Project has used PRA input value failure probability estimates principally to compute the CPAS contribution to Orion LOC. They have been used secondarily to help derive development test plans.

F-6. CxP and CEV Project decision makers who have made risk-informed design decisions based on CPAS LOC estimates were not fully informed of the source of the underlying data used to generate them. Top-level LOC uncertainties were communicated, but their method of derivation was not shown.

7.2.3 Requirements, Testing, Analyses, and Verification Planning

7.2.3.1 Requirements

The first draft version of the CPAS PTRS, dated November 2008, was provided at IDR-1 in December 2008 [Ref. 7]. AS had received it only 2 months previously. During the NESC assessment period the PTRS was revised once in May 2009. A more substantial revision was still in progress at the end of the assessment in April 2010. The Project indicated that interim NESC recommendations targeted to specific requirements would be incorporated into this version prior to PDR.

At a top level, the assessment team determined that the PTRS (Rev. A) [Ref. 7] did not define pass/fail criteria quantitatively. Moreover, it did not require an explicit statement of test suitability, assumptions, and/or statistical confidence in the test and/or simulation data and analysis. Thus, the PTRS was ambiguous and did not support defining verification test resources.

In colloquial terms, the PTRS did not state what needs to be known, how well it needs to be known, or how it will become known that it has been learned for relevant performance metrics necessary to recommend the system for flight operations.

For example, the format of the descent rate requirement could be stated as, "The CPAS will ensure with XX percent statistical power at an effect size of $X.X$ m/s that the CM will not exceed 10.7 m/s." Another example related to component strength might be stated as, "The CPAS will ensure with 90-percent confidence that the 0.3-percentile (3 sigma) of the strength of component X exceeds X ft-lbs."

These two examples support rigorously estimating test resources and provide a clear pass/fail metric to determine when a test objective has been met.

F-7. The PTRS does not define quantitative pass/fail criteria, nor does it require an explicit statement of test suitability, assumptions, or statistical confidence in the test and/or simulation data and analysis.

7.2.3.2 Development Test Plan

The assessment team review of the CPAS Gen-1 test reports suggested that in 2007 and 2008 there was no systematic process to establish the basis for and the conduct of testing. The Apollo Program Experience Report–ELS, NASA TN D-7437, (1973), illustrated the great extent of testing that was necessary to demonstrate the overall ELS system reliability to verify its suitability for manned flight. It also illustrated the need for thorough systems integration and compatibility verification. The ELS Project adopted a comprehensive systems engineering approach to design, development, and testing to deliver a robust, reliable system.

The assessment team reviewed the draft development test plan during RP-2. It appeared to be a parachute-centric test plan, a continuation of the Gen-1 testing. The assessment team decided that it was not a complete plan for demonstrating reliability of an integrated CPAS. The tests in the matrix did not appear to have been derived with consideration of a functional decomposition of CPAS parts. The tests showed an incomplete consideration of integrated performance of CPAS with other LRS components or with other CM subsystems.

The test objectives in the CPAS development test matrix were derived principally from *perceived or assumed* system-level risks. Risks in the development test matrix loosely but not completely reflected risks delineate in the constantly evolving FTAs and PRAs. The assessment team observed that the risk severities used to rank test objectives were determined by consensus of the CPAS Project development test team.

A risk-based testing approach has been successfully used during *operation* of the SSP and air transport systems to continuously improve system reliability and safety. However, a number of issues can arise when applying this selective approach during a developmental program that has little to no recent experience or flight data. It can be difficult to obtain pertinent failure data to formulate an accurate risk profile and there is a chance of overestimating or underestimating a given risk. A top-down, risk-based testing approach also can fail to identify developmental problems associated with process, material, workmanship, design errors, and interfaces. These can seriously threaten system reliability and can adversely impact development schedules.

Flight, ground, and subscale tests in the matrix were developed from the risk-based objectives. Flight-test configurations were created from parameters required to create test conditions (e.g., dynamic pressure, altitude, wing loading, wake, whether a representative parachute bay or FBC was required, etc.). Unique test configurations were identified, test objectives were grouped, and notional test vehicles were assigned. The flight test plan included main parachute performance evaluations of suspension line ratios, riser lengths, over-inflation control lines, and canopy porosity. Sub-scale test series were designed to assess risks related to CPAS separation from the PTV. The matrix included 16 drop tests in various configurations and at various altitudes, and nine ground tests.

The assessment team's view was that the priority of a parachute developmental test program should be to obtain as much data and learn the strengths and weaknesses of various design options as quickly as possible without being subject to the rigors and controls associated with performing qualification or flight certification tests. Concept feasibility tests should be conducted at the lowest level of sub-components possible. As the architectural elements of a Project mature and requirements are better defined, more sophistication tests then can be conducted.

The assessment team concluded that the CPAS development test plan had not been designed comprehensively and may not have been sufficient to obtain the required knowledge of system performance to mitigate the design risks. While the plan was derived from *perceived* or *assumed* risks as assessed by the Project, it did not link to *all* risks and concerns identified by FTA, FMEA, and PRAs. Some risks were sequestered to an 'off-scope' status. To adequately address these risks, priority should be given to those system elements identified by the FTA to have the highest uncertainties, thereby performing "risk balancing." The scope should include component-level verification tests to track compliance with requirements and mitigate or accept risks. Furthermore, the plan should provide V&V of simulation models, which are relied on for performance estimation in the absence of physical test data.

The assessment team assessed that the development test plan lacked precise objectives and success metrics that enabled clear decisions regarding when test objectives had been met. Uncertainty in performance metrics requires replication to assess experimental test variability. While some system parameters were replicated in the development plan, it exhibited confounding of multiple factors making it difficult to distinguish individual factor effects. The development test matrix represented a fraction of all possible combinations, constrained by practicality, but it did not represent a strategic fraction derived to meet precise objectives.

Given resource limitations and the complexity and highly integrated nature of CPAS, traditional, one-factor-at-a-time development testing may insufficiently assess performance factors, environmental factors, and interface interactions. All of these can affect system reliability. The assessment team concluded that a statistics-based DOE approach should be considered. DOE could provide an integrated, factorial test design to strategically and comprehensively plan the CPAS development tests. Such an approach would at a minimum include defining: (1) precise, quantifiable test objectives; (2) required statistical confidence to support test results, linked to risks and consequences; and (3) mathematical justification of test methodology and resources to support objectives and provide required confidence.

The DOE approach would also assist identifying what tests could be conducted for model validation, sub-system verification, and full-system verification and qualification.

F-8. The CPAS 'one-factor-at-a-time' development test program will be less efficient in assessing performance, environmental factors, and interface interactions that could affect system reliability than could be possible using statistics-based DOE techniques.

7.2.3.3 Ground Testing

The assessment team determined that the test plan placed an inappropriately low emphasis on ground testing as an effective means for understanding and assessing top risks early in the development phase. The assessment team concluded that four ground test programs were particularly needed at this time to mitigate risks associated with the following key events in the CPAS Con Ops:

1. Main parachute pack retention and release.
2. FBC panel jettison and trajectory.
3. Pilot parachute interference or contact with a panel during deployment following an abort.
4. Drogue parachute interference or contact with a panel during deployment following a nominal reentry.

7.2.3.4 Main Pack Retention System

The team assessed reliability risks with the main parachute pack retention system as it evolved throughout the assessment. At the end of RP-3 the system relied on a series of daisy-chain loops to retain the main parachute d-bag to CM gussets and to react flight loads. While this approach has been used successfully in other systems, including the Apollo Program, the assessment team thought that for CPAS the proposed method exhibited rigging complexity that introduced new risks of failed or delayed extraction of a main parachute pack.

As described in Section 6.3.3, the retention system design would incorporate flaps that restrained the main parachute packs along the side edges and interfaced to the segment gusset walls. The mechanism by which two pilot riser halves would impart forces to unzip their associated daisy-chain loops appeared immature. The riser halves would require one-for-one integration with each locking loop to pull each loop through its associated grommet. There was an implied assistance from the pilot chute risers to assist in the unzipping of the daisy chain without details. This concept had no apparent precedent and introduced development risk.

There had been no demonstration of this proposed release method and no previous example shown. The team assessed that there was substantial risk of abrasion and heat damage to the daisy-chain assembly, which could result in either failure to release or premature release of a main pack. Either of these could result in a total main parachute malfunction. The assessment team concluded that these risks should be retired by a series of demonstrations of the retention release function prior to in-flight full-system testing. These tests could include the addition of abraded or thermally damaged daisy-chain loops, to fully evaluate the robustness of the design.

At the end of RP-3, the main parachute d-bag concept contained a system of corsets and beckets on the outer surface to support the convex-curved upper-outer edge of the main parachute pack. As the main parachute pack density increased during development, it became more difficult for the parachute retention system to retain the outer surface shape of the d-bag through the total mission environmental cycle. The assessment team assessed that it will be imperative for the retention system to maintain the clearance between the outer surface of the d-bag and the inner surface of the FBC. Inertial and vibration forces may shift a pack in its retention system, altering its top and outer surfaces and particularly affecting its curved upper-outer edge. Such movement or distortion could allow contact of the d-bag with the FBC IML, consequently damaging parachute components or the airbags that jettison the FBC panels.

The assessment team noted that the CxP Ares 1-X recovery parachute failure investigators determined that a contributing cause of that failure was premature activation of a first stage parachute reefing line cutter, possibly from excessive movement in the parachute pack during flight prior to deployment. Ground tests can help ensure the adequacy of a main cutter actuation system against design load safety margins, and this can preclude premature reefing line cutter initiation during packing, transport, or flight, while ensuring actuation during the dynamic conditions of deployment. Cutter design should include a method to detect a premature firing at least to the point of installation.

The assessment team concluded that the Project's planned ground test was insufficient to address the identified risks. The Project should conduct ground tests to assess the viability of the main pack retention system early in the development effort, using a representative parachute compartment and parachute system. Testing should include ground static, thermal, inertial, shock, and vibration loads as part to any evaluation. The ability of the retention system to securely hold the CPAS components in place should be evaluated during these tests, including but not limited to, preventing gross movement and contact with the FBC IML. The effect on internal components such as the reefing line cutters should also be evaluated.

F-9.   The ground test portion of the Project's development test matrix insufficiently addresses risks associated with the main parachute pack retention and release system.

7.2.3.5 Forward Bay Cover

The assessment team assessed that the Project decision to segment the FBC into six airbag-jettisoned panels was made with limited consideration of key consequential aspects. Among these were effectiveness of the multiple panel design, fail-safe requirement, airbag load reaction on the packed main parachutes, panel separation line features, volumetric requirements, and panel release velocity uncertainties. The assessment team observed that multi-body trajectory analyses were not conducted of dispersed relative trajectories of the panels, drogue chutes, pilot chutes, main chutes, or other liberated components.

The team assessed that the change from a monolithic FBC to the six-panel configuration introduced additional complexity and failure modes. Project analyses of the risks and consequences of unsuccessful panel jettison due to either MDC or airbag failure were not shown. A failure of one or more of the panels to jettison may result in the improper deployment; failure to deploy; or damage to the drogue, pilot, or main parachutes or to associated risers, bridles, or CMUS components. It was not shown if retained panels were designed to break away by drogue or pilot parachute deployment as a back-up mode of jettison.

The team assessed that the jettison airbag system that was intended to thrust the individual panels away from the CM imposed additional complexity and sources of risk. One Project briefing indicated that each airbag would exert 32 psi over an area of 59 in^2 on the face of the main parachute pack. This would result in a total force of almost 1,900 pounds.

Airbag forces applied against the face of a main parachute pack assembly would represent a hazard to the deployment bag, its contents, and its retention system, possibly leading to damage to the main parachute components or premature release by rupturing the securing system. Airbag forces applied against the face of the drogue mortars could cause failure of the cover shear pins, leading to premature release of the drogue parachute. The addition of an airbag push-off structure in the sixth forward bay sector may introduce riser entanglement risks.

The use of individual airbags required interconnection to redundant pressure source for inflation, and redundant initiation. This multiplied the source of risk of failure. The airbags were shown to be attached directly to the panels, putting them at risk of thermal damage during reentry. As of the end of RP-3, no ground tests had been conducted to demonstrate the function or robustness of the panel jettison design.

The assessment team assessed that the risk and consequences of near-field contact and far-field contact of the FBC panels and parachutes had not been rigorously considered. Initial trajectories and motions of panels would be sensitive to variations inherent in several closely sequenced initial events, including MDC detonation events, initiation of airbag inflation, rate of airbag inflation, and push-off force.

During descent after reentry, a short-time interval was planned between FBC panels' jettison and drogue mortars' firing to reduce the risk of CM angular-rate buildup. The assessment team considered that this may introduce the risk of a drogue parachute being fired into and contacting a jettisoned panel. Following an abort, a short-time interval was planned between FBC panels' jettison and pilot mortars' firing to reduce the risk of CM angular-rate buildup. The assessment team considered that this might introduce a risk of a pilot parachute being fired into and contacting a jettisoned panel.

The assessment team concluded that panel jettison airbags must provide positive separation of *all six* panels from the CM, driving them outside the deployment radius of any of the parachutes and setting them on flight paths that would not intercept the CM or its parachutes. Jettison might occur in a complex air stream that was highly dependent upon capsule orientation relative to the wind vector. Based on part count, the probability of any one jettisoned panel contacting a parachute might be greater than the probability of a single monolithic cover contacting a parachute.

The team determined that panel ballistic coefficients could exhibit a wide range and that this could require a complex analyses and certification program that might introduce schedule risks.

The assessment team concluded that early in the development effort, the Project should validate the aerodynamic trajectory analysis by conducting a series of appropriate wind-tunnel, ground, and flight tests. These may include subjecting a family of test panels (ones that envelope a "best-guess" range of panel dimensions) to vertical wind-tunnel testing or free fall from an aircraft as well as investigating panels with variously deflated jettison airbags in this same manner. This would allow the characterization of the panels' drag in various stability modes. This data could then be coupled with the variation of the airbag thrust vector and airflow around the CM at off-nominal attitudes to develop full, 3D, multi-body, Monte Carlo trajectory analyses of the FBC panels relative to the CM and various parachute components. The resulting dispersion patterns of the FBC panels, CM, and parachute components could then be more comprehensively assessed.

7.2.3.6 Parachute Near-Field Contact during Deployment

The assessment team observed that the forward bay sectors in which each main pack was stowed were geometrically more restrictive than were the Apollo Program ELS stowage bays. This was reasoned to increase the risk (relative to the ELS) of extraction hindrance. The CPAS forward bay gusset geometry provided a 60-degree sweep of unhindered extraction, compared to 90 degrees for the ELS. In addition, the ramp on the upper back side of the three CPAS main d-bags would hinder upward extraction above the angle of the ramp. Main pack contact with CM structure during extraction could result in a failure to be extracted, a delay in extraction, or damage during extraction.

The team assessed that the previously noted low-fidelity characterization of FBC panel trajectories also posed a risk following a nominal reentry, of near-field contact of a drogue parachute fired directly into a recently released panel that had insufficiently cleared the space in front of the mortar. This same risk may exist following an abort, for a pilot parachute fired directly into a released panel that had insufficiently cleared the space in front of the mortar.

The team assessed that a series of ground tests should be conducted early in the development effort to demonstrate and assess the viability of extracting CPAS components. This testing should include ground static loads, inertia loads, shock, and vibration conditioning as described in Section 7.1.3. The extraction tests should include a series of static and dynamic strip tests where the d-bag is pulled at various angles on a representative forward bay mock-up. The pull angle selected should attempt to induce the worse possible interference or snag hazard to the d-bag as it is extracted. The d-bag should be extracted out and dragged across the outer surface of the CM at all possible angles. Snag hazards should be identified and documented as to their severity and risk. Relative extraction risks associated with geometric differences between CPAS and ELS should be assessed if ELS test data are applied in the CPAS PRA.

F-10. The ground test portion of the Project's development test matrix insufficiently addresses risks associated with the FBC panel jettison system and trajectories of released panels.

F-11. The ground test portion of the Project's development test matrix insufficiently addresses risks associated with near-field contact or interference of main parachutes with CM structure during deployment.

F-12. The ground test portion of the Project's development test matrix insufficiently addresses risks associated with near-field contact or interference of pilot or drogue parachutes with FBC panels during deployment.

7.2.3.7 Modeling and Simulation

The key analyses reviewed during the NESC CPAS independent review were taken primarily from the IDR, IDAT products, and various meetings, including CPAS SE&I, Hardware IPT, and MVP meetings.

The use of DCLDYN as the primary simulation tool for CPAS implied that it had undergone a V&V to ensure accurate modeling of the various trajectory phases. V&V approaches typically require extensive ground and flight tests to collect data to verify simulation accuracy. The test data collected and its associated uncertainty must be considered when formulating a model to simulate performance. Since flight tests cannot cover every expected flight regime, simulation is a key element in verifying any system across its expected operating ranges for flight. The assessment team determined that DCLDYN validation regarding its use in design decisions would be crucial in CPAS system design for its operational phases.

DCLDYN used Monte Carlo simulation to generate CPAS system results across various operating conditions including pad abort, ascent abort, and nominal re-entry. During the IDR meetings, DCLDYN was described as a two-body simulation that allows for three degrees of freedom (DOF) of each body in the vertical, horizontal, and pitch plane. The two bodies are connected by a non-linear spring, which represents the parachute structure. This model's validity was not apparent during the IDR regarding past use and verification with other model developers (i.e., verifying that the Orion aero-database was correctly implemented). The assessment team assessed that the validation of the DCLDYN simulation tool with regard to accurate performance predictions was a key missing element.

Reviewing other simulations being used in Orion development, the common number of DOF per body are three translational and three rotational in an inertial reference frame (six DOF). Rotational DOF are important for modeling capsule aerodynamic effects as well as the dynamic interactions between the capsule and parachute system. Techniques and assumptions used in building DCLDYN models account for only two translational DOF and one rotational DOF. From the IDR-3 presentation charts, the drogue design loads were seen to increase due to vehicle dynamics modeling in DCLDYN. Applicability of Orion models to DCLDYN therefore needs to be reviewed such that key characteristics and limitations of the missing DOF are understood and conveyed to decision makers.

At the beginning of the NESC's review, it was not recognized whether there was an independent V&V effort for DCLDYN. During the time of the review, another simulation called DSS was seen to predict parachute performance well for the Gen-1 testing. By the end of the review, an independent V&V role was seen to occur between DSS and DCLDYN. This addressed a key interim recommendation made near the beginning of the NESC review to use validated models.

Other analysis tools being used for CPAS design should also be V&V for their respective uses. For example, it was not clear that the tool used for an IDAT analysis of FBC segments far-field re-contact was valid for the type of study conducted. Specific assumptions and Monte Carlo parameters were not identified. The NASA Standard 7009, "Standard for Models and Simulations," would help the Project to V&V their tools for specific studies and help inform the responsible decision makers of results applicability and associated uncertainty [see Reference 10].

7.2.3.8 Uncertainty Analysis of Simulation Results

The modeling parameters used in DCLDYN for the parachute Monte Carlo analysis directly affect the design loads used to design and construct CPAS. The PTRS indicated that parachute loads were to be limited to a specified sigma level (meaning, by inference, variation about a mean value). An appropriate definition of "sigma" was missing from the PTRS and was therefore left open to interpretation. From the DCLDYN results, only a sigma number, which was thought to assume a normal distribution, was presented at the IDR as the design-to-number for the parachute loads. No presentation of the type of resultant distribution (e.g., histograms, quantile-to-quantile, cumulative distribution functions, etc.) was given from the Monte Carlo data, nor was a confidence level associated with that result. From the IDR slides, uniformly distributed numbers were used for the inputs, therefore implying the possibility of uniformly distributed outputs. The team assessed that a sigma number (assuming normally distributed data) would not be adequate to characterize the parachute load performance due to the inputs used and that more statistical rigor would be needed to better understand the results in support of decision making.

7.2.3.9 Clustered Parachute Modeling

The CPAS parachute system is composed of a set of drogue, pilot, and main parachutes working as a clustered system. Uncertainty in deployment, inflation, and other environmental factors can cause different opening loads between parachutes. Understanding the load sharing is necessary in designing margin into the design.

The parachute model employed by DCLDYN does not allow modeling of multiple parachutes. Instead a single parachute model must be contrived to have the same full open drag area and representative opening times as any desired cluster. The partitioning of the load between individual chutes in the cluster is then estimated through the use of an assumed load-sharing factor. For the CPAS design this has been derived from a limited set of past empirical data. This load-sharing factor was *derived* from Apollo Program test data and is a ratio of 65/35 for two parachutes and 50/25/25 for three parachutes. Using this load-share factor, the design limit loads for an individual parachute were specified to be greater for a two-chute drogue deploy than a single drogue deploy (34.2 versus 24.1 klbf) and similar for the three-chute mains (34.8 klbf). The team assessed that this load-sharing approach may be driving excessive conservatism in the design and could be better understood by developing a multi-body simulation that can model individual parachutes in a cluster. By developing this capability, the load share can be better understood and applied to the design effort.

F-13. CxP and CEV Project decision makers who have made risk-informed design decisions based on Project analysis results may not have recognized that key CPAS tools and models were not V&V for this application, nor may they have realized the amount of uncertainty in the analyses results.

7.2.3.10 Master Verification Plan

At the first IDR in December 2008, presentations were made on the embryonic work toward the development of a test and verification plan. By mid 2009, a MVP development was begun in earnest and a Project team was formed. The process employed to develop the MVP is shown in Figure 6.5-3. In October 2009, the Project MVP team augmented its membership to include a statistician trained in statistical DOE. MVP development continued until December 2009, at which time the MVP process was reduced to a low priority to focus on the PTRS revision. The re-prioritized focus on the PTRS revision emphasized the vital inter-relationship between the MVP and PTRS.

The Project generated a V&V document to address the Project's system integration needs and requirements. This document was intended to provide minimum acceptable verification success criteria, general information concerning the verification methods and resource requirements, and a link to the final verification evidence. The CPAS Certification Plan was used to define the detailed tools and resources, schedules, and inter-relationships needed to support successful system certification. The V&V matrix referenced in the V&V document and generated for PDR was the typical matrix in spreadsheet form showing success criteria and verification method. However, this spreadsheet lacked the fidelity and interconnectivity required for such a large and complex system as CPAS.

The Project's answer to the complex integration issue was to develop a MVP that captured the individual VA. Each requirement addressed in the PTRS may generate multiple VAs. VAs may generate multiple analyses and tests. A VA concept was authored by the Project to address the issues: types of analyses required, where the information to support these analyses would come from (e.g., test, inspections, etc.), identification of other activities supported by the same analysis, and the different analyses needed to be performed. This VA concept identified the relationships between activities, enabled better scheduling and planning of analysis and testing resources, and allowed for impact assessments for changes or delays caused by predecessor activities. The Project recognized that the verification plan must be started early in the design process and be complete by the Critical Design Review. The Project also recognized that the VA concept should be extended to include development testing and analysis, not just verification tests and analyses.

The CPAS SE&I team presented the MVP concept to the IPTs and to CEV Project Management. Expectations (ground rules) for the VAs, roles and responsibilities of participating organizations and work groups, and the review/approval process were discussed during a kickoff meeting. VAs were developed using the following hierarchy:

1. CPAS PTRS is the parent document.

2. CPAS V&V document defines the success criteria from the PTRS as Verification Requirements (VRs).

3. VAs are derived from VRs. Each VR will have at least one VA. However, in many cases, each VR will produce multiple VAs. A VA can produce other VAs to attain missing/insufficient data to complete the parent VA.

While a VA is under review by the IPTs, System Integration, and CEV Project Management, it is considered a candidate VA. Once approved by the stakeholders, the VA is moved to "accepted" and entered into CRADLE™.

The assessment team determined that while the MVP process to trace test requirements to the PTRS was rigorous, a lack of clarity in the PTRS (previously discussed) resulted in an inordinate use of resources discussing precisely what would satisfy PTRS requirements. In addition, there appeared to be a disproportionate focus on system-level flight tests, due to the cost and schedule constraints, rather than a systematic approach to integrate, and exploit, ground-testing capabilities.

As "a chain is only as strong as its weakest link," CPAS system reliability and crew safety is dictated by its weakest building block or component. The most important goal of CPAS verification, or any verification, should be finding the weakest link or potential weak links from the many building blocks and processes aggregated into a single system.

Based on reviewing CPAS testing and verification planning documents provided, it was clear the Project sought to achieve this goal by utilizing systems tools to identify and selectively mitigate the weak links based on their risk outcome(s). However, this targeted verification approach is most effective when it is used in parallel with a component-based verification plan that covers system components. The central theme was to create a complete CPAS verification data picture of a functional verification matrix to effectively manage the VA while avoiding any weak or missing links and eliminating any unnecessary duplication.

An example of the functional verification matrix and possible contents in the matrix are described as:

- Systematically decompose the CPAS subsystem into its components and their functions, including *all* functions that lead to factors which can influence CPAS performance.

- Identify interface components (from other subsystems) that may influence CPAS performance, and their functions.

- Identify environmental factors that can influence component performance.

- Identify any unique handling, processing, and inspection requirements.

- Include *all* risks and concerns identified by FTA, FMEA, and PRA.

- Derive verification parameters traceable to requirements, risks, and operations concepts.

- Derive an inclusive list of candidate verification tasks (analyses, tests, inspections, and demonstrations) that rigorously trace back to risk analyses, component functional analyses, requirements, and operations concepts.

- Cross check the existing development plan and the draft verification plan to reveal potentially unidentified verification tasks that will be necessary to certify the system for human-rated flight.

In summary, the assessment team determined that the draft MVP appeared to emphasize aerial drop tests over ground tests. Instead of a bottoms-up test and verification approach, CPAS was more focused on system-level drop tests. The assessment team found that system-level drop tests represented a small proportion of the full suite of testing capabilities that could be leveraged to achieve successful verification. VAs should be performed at the lowest level possible and built up to mitigate risks and control external effects. For example, the riser assembly mechanical strength qualification can be done as a component-level test versus during an all-up system drop test.

For the MVP, the assessment team determined that a test matrix based on the most probable worst-case flight conditions and environmental factors (on a bad day/worst case) should be developed. The MVP should state these conditions in reporting results and flight should be restricted to those conditions which have been validated and verified. While the Project seeks to "test as you fly and fly as you test," the assessment team recognized that it is often impractical to develop a test matrix to encompass *all* external environmental variables and interactions between its sub-assembly components, especially for a system as complex as CPAS. Under these circumstances, DOE can be employed to systematically sample the multi-dimensional test space that is both broad and deep. In addition, DOE seeks to maximize the amount of information obtained from the entire test program, rather than sub-optimizing for individual tests. Due to the resource constraints on CPAS, the assessment team recommended employing DOE in all testing levels (component to flight) to optimally allocate resources for requirement verification.

F-14. The draft MVP was appropriately immature during the pre-PDR period of assessment. The draft underemphasized ground tests.

O-4. Verification activities should be performed at the lowest level possible and built up to mitigate risks and control external effects. A verification plan that is not comprehensively derived from a full inventory of component functional requirements may fail to identify failure modes that were not observed on heritage hardware or on previous tests on development hardware.

O-5. A verification matrix that is based on the most probable worst-case flight conditions and environmental factors (worst case/bad day) is ideal, but would be impractical for a verification matrix of a system as complex as CPAS to encompass external environmental variables and interactions between its sub-assembly components. Statistical DOE techniques can systematically sample a multi-dimensional test space in a manner that is both broad and deep, and can maximize the amount of information obtained from an entire test program, rather than sub-optimizing individual tests.

7.2.4 Organizational Complexity Impact on Systems Integration

The assessment team assessed that the structure of the organization of which the Project was a part, may pose integration risks that adversely impact CPAS reliability. The partition of Orion hardware and the contractual roles and responsibilities are described in Section 6.6.

The Project was responsible for drogue, pilot, and main parachutes; their suspension lines, risers, and deployment bags; confluence fittings (if any); the torque reducing keeper; and other ancillary components. The Orion Project's lead contractor, LM, was responsible for several other components critical to CPAS reliability, including drogue and pilot mortars, riser cutters, the fairlead fitting, and auxiliary parachute guns (if any). LM was also responsible for the integration of CPAS into the larger LRS, which included the CMUS. LM was also responsible for integrating the LRS into the CM and for module features that were critical for CPAS reliability, including the FBC and its jettison system, and the forward bay gussets and tunnel.

The assessment team's understanding was that design development and test data were to be provided by AS to the Project, through its contract with Jacobs ESCG. The parachute system would ultimately be provided by NASA to LM as GFE. AS did not have a production contract for CPAS at the time of this review.

The team assessed that AS brought valuable expertise to the Project. However, the assessment team thought that the complexity of the organization structure complicated the definition and ease of change of subsystems interface control documents (ICDs) and that certifying the integrated LRS would be less cost effective. Most significantly, the team assessed that the complex reporting lines, roles, and responsibilities arising from CPAS being GFE might pose risks to designing a well-integrated and reliable subsystem.

During the Apollo Program, the ELS organization was responsible for the total upper deck ICD, allowing control of anything that could affect or interfere with parachute function.

The team assessed that the unconventional separation of the CPAS drogue and pilot mortars from CPAS introduced a significant integration risk. Successful drogue and pilot parachute deployment will depend on key elements of mortar design, and inevitable changes in parachute design during development may critically affect mortar performance. Historically, single entity/design organization were employed for development of human-rated landing recovery systems to more readily accommodate changes in either the mortars or the parachutes and to facilitate a more efficient development and certification testing effort.

The assessment team thought throughout the assessment period that the organizational complexity introduced risks through the miscommunication of requirements or other critical design information. These risks could be mitigated by aggressive application of systems engineering best practices. One requirement of systems engineering is an organizational breakdown structure with defined roles and responsibilities.

The assessment team also was concerned that the complexity of the reporting structure (AS-Jacobs ESCG-NASA-LM) and the complex systems of boards and panels (GEMCP, ERB, etc.) with varying authority to approve or advise would introduce additional integration risks and hinder progress during development, validation, and certification.

The assessment team believed that the design of a robust and reliable CPAS and its integration into LRS would require ongoing discussions and iterative design trades with other CM subsystems throughout the development process. The team assessed early in the assessment period that CPAS would benefit from the creation of a Phase Lead or a Phase Team who would report to the SE&I team chair, with the principal responsibility of assessing CPAS component interfaces with other systems. An interim NESC recommendation in April 2009 was for such an entity to

> "Have broad-reaching authority to assess design changes for their effect on other systems that interface with CPAS, and for the effect those other systems have on CPAS, regardless of the interfaced system ownership. This (lead or team) should also coordinate the review and distribution of changes by the appropriate Technical Authority (TA) for each affected organization." {Interim R-9, Appendix B}.

The assessment team felt that the creation of the IDAT in June 2009 addressed this recommendation. The success of the IDAT in resolving several CPAS integration issues proved its effectiveness.

F-15. The nontraditional roles and responsibilities for development of components that are critical to reliable function of CPAS, but are not a part of CPAS, poses integration and communication challenges that may introduce unrecognized reliability risks to CPAS.

O-6. Risks arising from the nontraditional division of roles and responsibilities between the CPAS Project, LM, AS, and Jacobs ESCG have not been identified.

O-7. Concerted application of systems engineering best practices will be essential to reduce the increased risk posed by the nontraditional division of roles and responsibilities between the CPAS Project, LM, AS, and Jacobs ESCG.

After the IDAT was disbanded, the Orion Project formed the ELS-FIT with a charter to ensure that future design decisions will reflect integrated solutions. The assessment team was encouraged by this action which indicated that the risks associated with organizational complexity were generally recognized by the Project and that systems engineering practices would be employed in an effort to manage those risks.

The assessment team recognized that the Orion Project's multi-element organization was unlikely to be contractually simplified. The team therefore created an Agency lesson learned for future NASA space vehicle development programs with human-rated parachute recovery systems.

LL-1. NASA space vehicle development programs with human-rated parachute recovery systems should make a single organization fully responsible and accountable for the design, development, integration, and certification (DDI&C) of the deceleration and landing systems. This responsibility should encompass all participating subsystems that can affect the reliability of the integrated descent and landing system. Such an approach to DDI&C could have utility for other tightly coupled spacecraft systems of systems on which crew lives are critically dependant on reliable function.

8.0 Findings, Observations, and NESC Recommendations

8.1 Findings

The following *final* findings were identified:

F-1. The CPAS Project has not implemented a formal process to manage volume growth of components in the CM forward bay during design, development, evaluation, and operation.

F-2. Deployment of a torque reducer keeper on the CPAS main risers has no design precedent and poses risks to CPAS reliability. Extensive testing will be necessary to quantify these risks and demonstrate reliable functionality.

F-3. Using the CM RCS to orient the CM into a toe-forward position just prior to water landing is unprecedented and is an integrated issue that introduces risks to CPAS and the CM structure.

F-4. Failure probability estimate data derived from the 2009 EE exercise exhibited a large variance, greatly limiting its usefulness for generating CPAS PRA input values.

F-5. Failure probability estimate data derived from the 2009 EE exercise are invalid for the Gen-2 architecture.

F-6. CxP and CEV Project decision makers who have made risk-informed design decisions based on CPAS LOC estimates were not fully informed of the source of the underlying data used to generate them. Top-level LOC uncertainties were communicated but their method of derivation was not shown.

F-7. The PTRS does not define quantitative pass/fail criteria, nor does it require an explicit statement of test suitability, assumptions, or statistical confidence in the test and/or simulation data and analysis.

F-8. The CPAS 'one-factor-at-a-time' development test program will be less efficient in assessing performance, environmental factors, and interface interactions that can affect system reliability than could be possible using statistics-based DOE techniques.

F-9. The ground test portion of the Project's development test matrix insufficiently addresses risks associated with the main parachute pack retention and release system.

F-10. The ground test portion of the Project's development test matrix insufficiently addresses risks associated with the FBC panel jettison system and trajectories of released panels.

F-11. The ground test portion of the Project's development test matrix insufficiently addresses risks associated with near-field contact or interference of main parachutes with CM structure during deployment.

F-12. The ground test portion of the Project's development test matrix insufficiently addresses risks associated with near-field contact or interference of pilot or drogue parachutes with FBC panels during deployment.

F-13. CxP and CEV Project decision makers who have made risk-informed design decisions based on Project analysis results may not have recognized that key CPAS tools and models were not V&V for this application, nor may they have realized the amount of uncertainty in the analyses results.

F-14. The draft MVP was appropriately immature during the pre-PDR period of assessment. The draft underemphasized ground tests.

F-15. The nontraditional roles and responsibilities for development of components that are critical to reliable function of CPAS but are not a part of CPAS, pose integration and communication challenges that may introduce unrecognized reliability risks to CPAS.

8.2 Observations

The following *final* observations were identified:

O-1. A large cache of Apollo Program ELS test and reliability reports has been located and is being retrieved by the Project. Some of these data will be applicable for generating CPAS PRA input values because the ELS and the CPAS Gen-2 architectures are functionally similar.

O-2. Top-level system failure probability estimates often are estimated or predicted prior to PDR, but tests and analyses ultimately must be performed to validate and reduce PRA input value uncertainties.

O-3. The Project has used PRA input value failure probability estimates principally to compute the CPAS contribution to Orion LOC. They have been used secondarily to help derive development test plans.

O-4. Verification activities should be performed at the lowest level possible and built up to mitigate risks and control external effects. A verification plan that is not comprehensively derived from a full inventory of component functional requirements may fail to identify failure modes that were not observed on heritage hardware or on previous tests on development hardware.

O-5. A verification matrix that is based on the most probable worst-case flight conditions and environmental factors (worst case/bad day) is ideal but would be impractical for a verification matrix of a system as complex as CPAS to encompass external environmental variables and interactions between its sub-assembly components. Statistical DOE techniques can systematically sample a multi-dimensional test space in a manner that is both broad and deep and can maximize the amount of information obtained from an entire test program, rather than sub-optimizing individual tests.

O-6. Risks arising from the nontraditional division of roles and responsibilities between the CPAS Project, LM, AS, and Jacobs ESCG have not been identified.

O-7. Concerted application of systems engineering best practices will be essential to reducing the increased risk posed by the nontraditional division of roles and responsibilities between the CPAS Project, LM, AS, and Jacobs ESCG.

8.3 NESC Recommendations

The following *final* NESC recommendations are directed toward the CPAS Project unless otherwise indicated and were of the top-most concern to the assessment team at the conclusion of its assessment. The *final* NESC recommendations are all reiterations or enhanced restatements of *interim* NESC recommendations, which are noted parenthetically for reference. *Interim* NESC recommendations R-1 through R-30, their foundational findings and observations, and additional supporting material can be found in Appendix B. *Interim* NESC recommendations R-31 and R-32 can be found in Appendix C. *Interim* NESC recommendations R-33 through R-55 can be found in Appendix D.

R-1. Create a volume budget with volume margins for the CM forward bay; maintain it throughout the DDT&E effort; and continue to maintain it as design improvements are made after the CM becomes operational. *(F-1), {interim R-3}*

R-2. The Orion Project (or the delegated ELS-FIT) should conduct a comprehensive assessment of the overall feasibility and ramifications of using the CM RCS to perform the function of reorientation of the CM prior to water landing. *(F-2, F-3), {interim R-41}*

- Include representatives from CM subsystems (including CPAS) who interface with or are dependent on the RCS function.
- Include an assessment of LRS certification plans.

R-3. Discontinue using results from the 2009 expert elicitation exercise for generating PRA/LOC estimates until failure estimates are revised to reflect the current architecture and the cause of the large variability seen in the data from that exercise can be rectified; utilize applicable data from the geometrically similar Apollo Program ELS to generate estimates for applicable failure estimates or conduct tests and analyses to generate new PRA data. *(F-4, F-5, O-1, O-2), {interim R-14, R-45, R-46}*

R-4. Convey to CxP and CEV Project decision makers the source of all data and an estimate of its uncertainty, when generating post-PDR PRA input value failure probability estimates to be used to show the CPAS contribution to Orion LOC. *(F-6, O-3), {interim R-6, R-42}*

R-5. Revise requirements in the PTRS to include verifiable pass/fail metrics. *(F-7), {interim R-53}*

R-6. Re-plan post-PDR development testing and analyses activities strategically and comprehensively. Consider using statistical DOE techniques to maximize efficient resource utilization. *(F-8), {interim R-5, R-31}*

R-7. Consider inclusion of the following ground tests early in the post-PDR development plan to obtain data on risks associated with key events in the CPAS Con Ops:

- Main parachute pack retention and release. *(F-9, F-14), {interim R-38, R-39}*
- FBC panel jettison and trajectory. *(F-10, F-14), {interim R-34, R-36}*
- Main parachute interference or contact with CM structure during extraction. *(F-11, F-14), {interim R-37}*
- Drogue parachute interference or contact with an FBC panel during deployment following nominal reentry. *(F-10, F-12, F-14), {interim R-10, R-33}*
- Pilot parachute interference or contact with an FBC panel during deployment following an abort. *(F-10, F-12, F-14), {interim R-10, R-33}*

R-8. Convey to CxP and CEV Project decision makers the maturity and validity of analytical tools, models, verification status, input data, limitations, uncertainties, and credibility, when the results are to be used to support design decisions. *(F-13), {interim R-51}*

R-9. Implement a comprehensive verification plan that is linked to requirements. Consider using statistical DOE techniques for test verification activities to maximize efficient resource utilization. Perform verification activities at the lowest level possible and build up to higher levels, to mitigate risks and control external effects. *(F-14, O-4, O-5), {interim R-32, R-54}*

9.0 Lessons Learned

The following Agency lessons learned are identified. Both are reiterations or enhanced restatements of *interim* Lessons Learned that can be found in original form in Appendices B and D:

LL-1. NASA Programs and Projects should ensure contractor development data is contractually required for delivery to the Government, is recorded in Agency documents (e.g., TMs, SPs, CMs, etc.), and is stored in Agency searchable databases to assure their availability for future programs. *{interim O-23}*

LL-2. NASA space vehicle development programs with human-rated parachute recovery systems should make a single organization fully responsible and accountable for the DDI&C of the deceleration and landing systems. This responsibility should encompass all participating subsystems that can affect the reliability of the integrated descent and landing system. *(F-15, O-6, O-7), {interim R-9}*

- Such an approach to DDI&C could have utility for other tightly coupled spacecraft systems of systems on which crew lives are critically dependant on reliable function.

10.0 Definition of Terms

Corrective Actions	Changes to design processes, work instructions, workmanship practices, training, inspections, tests, procedures, specifications, drawings, tools, equipment, facilities, resources, or material that result in preventing, minimizing, or limiting the potential for recurrence of a problem.
Finding	A conclusion based on facts established by the investigating authority.
Lessons Learned	Knowledge or understanding gained by experience. The experience may be positive, as in a successful test or mission, or negative, as in a mishap or failure. A lesson must be significant in that it has real or assumed impact on operations; valid in that it is factually and technically correct; and applicable in that it identifies a specific design, process, or decision that reduces or limits the potential for failures and mishaps, or reinforces a positive result.
Observation	A factor, event, or circumstance identified during the assessment that did not contribute to the problem, but if left uncorrected has the potential to cause a mishap, injury, or increase the severity should a mishap occur. Alternatively, an observation could be a positive acknowledgement of a Center/Program/Project/Organization's operational structure, tools, and/or support provided.
Problem	The subject of the independent technical assessment.
Proximate Cause	The event(s) that occurred, including any condition(s) that existed immediately before the undesired outcome, directly resulted in its occurrence and, if eliminated or modified, would have prevented the undesired outcome.

	NASA Engineering and Safety Center Technical Assessment Report	Document #: NESC-RP-08-00487	Version: 1.0
Title:	CEV Parachute Assembly System Independent Design Reliability Assessment		Page #: 106 of 109

Recommendation   An action identified by the NESC to correct a root cause or deficiency identified during the investigation. The recommendations may be used by the responsible Center/Program/Project/Organization in the preparation of a corrective action plan.

Root Cause   One of multiple factors (events, conditions, or organizational factors) that contributed to or created the proximate cause and subsequent undesired outcome and, if eliminated or modified, would have prevented the undesired outcome. Typically, multiple root causes contribute to an undesired outcome.

11.0 Acronyms List

ADS	Aerodynamic Decelerator System
AMA	Analytical Mechanics Associates, Inc.
AS	Airborne Systems
ATK	Alliant Techsystems, Inc.
BEI	Bay Engineering Innovations
CEV	Crew Exploration Vehicle
CM	Crew Module
CMUS	CM Uprighting System
Con Ops	Concept of Operations
COPV	Composite Overwrapped Pressure Vessel
CPAS	CEV Parachute Assembly System
CxP	Constellation Program
CY	Calendar Year
D-Bag	Deployment Bag
DDI&C	Design, Development, Integration, and Certification
DDT&E	Design, Development, Testing and Evaluation
DFMR	Design for Minimum Risk
DOE	Design of Experiment
DOF	Degree of Freedom
DSS	Deceleration System Simulation
EDL	Entry, Descent, and Landing
EE	Expert Elicitation
ELS	Earth Landing System
ELS-FIT	Earth Landing System-Functional Integration Team
EO	Experience-based Observations
ERB	Engineering Review Board
ESCG	Engineering and Science Contract Group
FAQ	Frequently Asked Question

FBC	Forward Bay Cover
FIT	Functional Integration Team
FMEA	Failure Modes and Effects Analysis
FS	Factual Statement
FT	Fault Tree
FTA	Fault Tree Analysis
GFE	Government-Furnished Equipment
GN&C	Guidance, Navigation, and Control
GRC	Glenn Research Center
GSFC	Goddard Space Flight Center
HA	Hazard Analysis
ICD	Interface Control Document
IDAT	Integrated Design Assessment Team
IDR	Internal Design Review
IML	Inner Mold Line
IPT	Integrated Product Team
JSC	Johnson Space Center
LAS	Launch Abort System
LaRC	Langley Research Center
LM	Lockheed Martin
LOC	Loss Of Crew
LRS	Landing Recovery System
MDC	Mild Detonating Cord
MSFC	Marshall Space Flight Center
MTSO	Management and Technical Support Office
MVP	Master Verification Plan
NAWS	Naval Air Weapons Station
NESC	NASA Engineering and Safety Center
NG	Northrop Grumman
NRB	NESC Review Board
NSI	NASA Standard Initiator
NVR	Northrop-Ventura Report
OML	Outer Mold Line
OSIrIS	Outdoor Scene and Infrared Image Simulation
PA	Pad Abort
PDR	Preliminary Design Review
PRA	Probabilistic Risk Assessment
PTRS	Project Technical Requirements Specification
PTV	Parachute Test Vehicle
RAESR	Risk Assessment Executive Summary Report
RCS	Reaction Control System

RP	Review Period
R&R	Retention and Release
SAIC	Science Applications International Corporation
SE&I	Systems Engineering and Integration
S&MA	Safety and Mission Assurance
SRB	Solid Rocket Booster
SSC	Stennis Space Center
SSP	Space Shuttle Program
TBD	To Be Decided
TA	Technical Authority
TPS	Thermal Protection System
U.S.	United States
VA	Verification Activities
V&V	Validation and Verification
VR	Verification Requirement

12.0 References

1. *DDT&E Considerations for Safe and Reliable Human Rated Spacecraft Systems*, NASA-TM-2008-215126, Vol. 1, Executive Summary (2007).
2. IDR-1 PowerPoint presentation document (December 2008).
3. IDR-2 PowerPoint presentation document (April 2009).
4. IDR-3 PowerPoint presentation document (September 2009).
5. Parachute Recovery Systems Design Manual, T. W. Knacke (1992).
6. *Apollo Experience Report–Earth Landing System (ELS)*, NASA TN D-7437 (1973).
7. *PTRS for the Crew Exploration (CEV) Parachute Assembly System (CPAS)*, JSC-63497, Baseline version (November 2008), Rev. A version (May 2009).
8. *Assumptions Companion Document to the PTRS for the Crew Exploration Vehicle (CEV) Parachute Assembly System (CPAS)*, JSC-64355, Rev. A version (October 2008), Rev. B version (May 2009).
9. *Expert Elicitation Summary Report for the Crew Exploration Vehicle (CEV) Parachute Assembly System (CPAS)*, JSC 64967, ESCG-6110-09-SS-DOC-0744 (2009).
10. *NASA Standards for Modeling and Simulation*, NASA-STD-7009 (2008).

Volume II: Appendices (Stand-alone Volume)

Appendix A. Stakeholder Request (November 2008)

Appendix B. Stakeholder Outbrief of Interim Recommendations 1 and supporting material (April 2009)

Appendix C. Stakeholder Outbrief of Interim NESC Recommendations 2 and supporting material (September 2009)

Appendix D. Stakeholder Outbrief of Interim NESC Recommendations 3 and supporting material (April 2010)

www.ingramcontent.com/pod-product-compliance
Lightning Source LLC
Chambersburg PA
CBHW081729170526
45167CB00009B/3755